THE
WASTE-FREE
WORLD

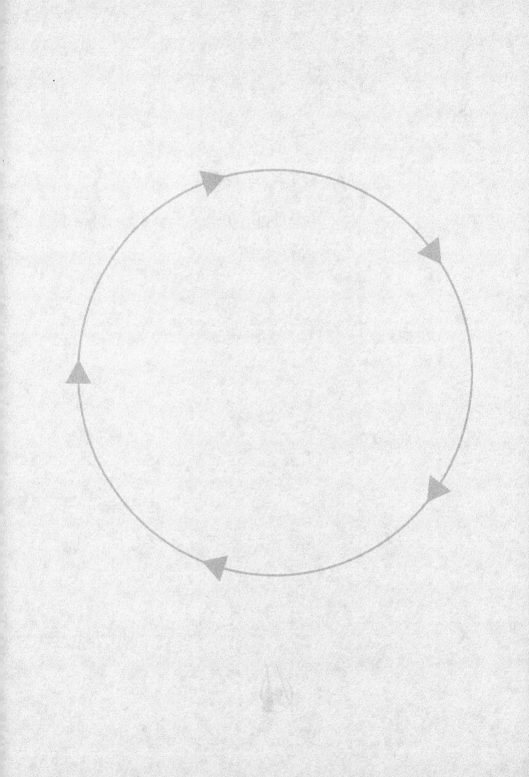

THE
WASTE-FREE
WORLD

**How the Circular Economy Will Take Less,
Make More, and Save the Planet**

RON GONEN

PORTFOLIO / PENGUIN

PORTFOLIO/PENGUIN

An imprint of Penguin Random House LLC

penguinrandomhouse.com

Most Portfolio books are available at a discount when purchased in quantity for sales promotions or corporate use. Special editions, which include personalized covers, excerpts, and corporate imprints, can be created when purchased in large quantities. For more information, please call (212) 572-2232 or e-mail specialmarkets@penguinrandomhouse.com. Your local bookstore can also assist with discounted bulk purchases using the Penguin Random House corporate Business-to-Business program. For assistance in locating a participating retailer, e-mail B2B@penguinrandomhouse.com.

p. xviii "Linear vs. Circular Economy," copyright © 2016 by Catherine Weetman, no changes made, via CC BY-SA 4.0 https://creativecommons.org/licenses/by-sa/4.0/legalcode; p. 21 "Throwaway Living," *Life* magazine, 1955. Photographer Peter Stackpole/ The LIFE Picture Collection via Getty Images; p. 26 "Environmental doughnut infographic," Designed for Kate Raworth, by DoughnutEconomics, no changes made, via CC BY-SA 4.0 https://creativecommons.org/licenses/by-sa/4.0/legalcode.

Library of Congress Cataloging-in-Publication Data

Names: Gonen, Ron, author.
Title: The waste-free world: how the circular economy will take less, make more, and save the planet / Ron Gonen.
Description: New York: Portfolio / Penguin, [2021]
Identifiers: LCCN 2020050212 (print) | LCCN 2020050213 (ebook) | ISBN 9780593191842 (hardcover) | ISBN 9780593191859 (ebook)
Subjects: LCSH: Industrial management—Environmental aspects. | Business enterprises—Environmental aspects. | Sustainable development.
Classification: LCC HD30.255 .G65 2021 (print) | LCC HD30.255 (ebook) | DDC 338/.064—dc23
LC record available at https://lccn.loc.gov/2020050212
LC ebook record available at https://lccn.loc.gov/2020050213

Paperback ISBN: 9780593853900

BOOK DESIGN BY TANYA MAIBORODA

147141878

I dedicate this book to my mom
who taught me about work ethic and dedication,
to Paul and Linda for their support,
and to my wife and children
for being my passion.

Contents

Introduction: Beyond
Sustainability to Renewal

WALKING ALONG A PATHWAY rimmed on both sides by fields of lush, knee-high green grass, I find myself in awe that I'm strolling on terrain once dominated by New York City's most appalling eyesore—perhaps even more offensive to the nose—the infamous Fresh Kills Landfill. It's recent transformation is a testament to the vigor with which nature can restore the beauty and ecological richness that had been stolen.

Once a vibrant wetland estuary named for its fresh streams, Fresh Kills was foraged in the early twentieth century by a thriving Italian immigrant community seeking wild mushrooms and grapes, dandelions and watercress, its waterways harvested for mud shrimp, steamer clams, and oysters. When the area was selected in 1947 for the dumping of virtually all of New York City's trash, some Staten Islanders were so alarmed that they proposed the island secede from the city. As feared, the 2,200 acres were transformed into four moldering mountains of foul refuse said to smell like rotting blue cheese. At 250 feet, they towered higher than the neighboring Statue of Liberty, comprising not only the world's biggest landfill but the largest man-made structure on the planet, their volume exceeding even that of the Great Wall of China's.

Finally, after decades of public outcry, the landfill was closed in 2001. New York City then launched a year-long restoration effort, covering vast swathes of refuse with layers of lining, topped with soil and planted with wild grasses, to transform Fresh Kills into a beautifully landscaped park featuring undulating wild-flower meadows, kayaking, and canoeing. A profusion of wildlife returned, from butterflies, bees, and bats to blue herons, sparrows, ospreys, owls, and large raptors called fish hawks. Even deer, coyote, and red foxes have ventured into the rapidly rewilding stretches of grassland and forest.

Rounding a bend in the path, I spot the dense foliage of what arborist William Bryant Logan has described as a forest engaged in a furious battle of renewal, with tangles of bittersweet and poison ivy vines tearing down red maple, mulberry, and black cherry strivers, from whose exposed roots new shoots fight their way toward the sky—a battle known by foresters, Logan explains, as "phoenix regeneration." Fresh Kills has become an exemplar of the ecological wizardry by which the natural world, when given the chance, can heal itself. As we enter a new decade, our industrialized human economy has the opportunity to follow nature's lead.

We've Been Scammed

How could the monstrosity of the Fresh Kills Landfill ever have seemed like a solution to the rapidly spiking volumes of refuse Americans began generating during the economic boom that followed World War II? Fresh Kills became the most ludicrous representation of a perverted form of capitalism that established dumping on publicly owned land as an inalienable right of major industries. Its mountains of garbage proliferated, in less glaring fashion, all around the U.S. as industrial and consumer goods giants shirked off responsibility for the horribly wasteful throwaway economy they hooked Americans on. They pass themselves off as free-market capitalists, but their profits are in fact heavily dependent on taxpayers funding the disposal of their products in landfills. Without taxpayers footing that bill—a modern form of socialism—their profits would be seriously diminished.

I learned firsthand just how absurd our wasteful system is in 2012, when I was appointed as the Deputy Commissioner for Sanitation, Recycling and Sustainability for New York City. I was tapped by Mayor Michael Bloomberg because I had cofounded a start-up in 2003, Recyclebank, that boosted recycling rates in cities all around the country. I was appalled by the deception perpetrated not only on New Yorkers but on the public all around the country—and on people all over the planet—by coalitions of waste profiteers. They've richly benefited from a take-make-waste economic system at the great expense of consumers, consumer goods companies, city and town governments, and the health of the planet. In New York City, over $300 million was being spent annually exporting waste to landfills in Pennsylvania, Ohio, and South Carolina. Most of it could have been recycled locally, avoiding this costly tax burden. That $300 million would be enough to nearly double the number of teachers or the number of police officers in the city. Or it could have been used for a tax cut without having adverse effects on municipal services. This literal waste of taxpayer money is repeated in almost every city in the United States. Even in cities with excellent recycling service and participation, far too much waste that could be eliminated with efficiencies in packaging or by being recycled locally is sent to landfills. The result is that for the past few decades, U.S. cities have spent billions on exporting waste that could have been used instead to improve infrastructure and social services.

At Recyclebank, our technology allowed us to offer points to households based on the volume of recyclables they put out in their bin, which could be used to shop at supermarkets, restaurants, and family entertainment spots. When we partnered with cities or individuals that signed up for our service, we explained that the public pays vast sums to landfill companies to haul and dump commodities like paper, metal, glass, and plastic that could be sold to recycling firms. And we showed that neighbors who *were* recycling were unknowingly footing some of the disposal bill for their neighbors who weren't. Why should anyone have to pay more in taxes just because some of their neighbors disposed of potential revenue?

When people realize the direct cost of their waste to their household, and the potential value of recycling it, behaviors change. We found that people hadn't understood the direct economic value of recycling because, as with green initiatives generally, recycling is so often disparaged as pointless do-gooderism. As we'll explore more fully later, that's propaganda from companies and industries with a vested financial interest in undermining recycling. Too often, the press has been a willing participant in perpetuating the fictitious story that environmental conservation and restoration are contrary to our economic interests.

The good news is that Americans currently recycle over 90 million tons of metal, paper, plastics, electronics, textiles, and glass annually. If they didn't, communities collectively would have to pay an additional $3 billion annually in landfill disposal fees. What's more, they would forgo the more than $100 billion in economic activity generated by the recycling industry in the U.S., including 540,000 American jobs. Recycling can also reduce the cost of the materials to make the products we buy, which can bring down prices. Recycling, in short, is a matter of economic self-interest.

For cities with advanced recycling infrastructure and strong local markets for recycled material, it's proved an economic boon. New York City, for example, has a long-term public-private partnership with Pratt Industries to convert all its wastepaper locally into new paper products that are sold back into the New York City market. The city is paid for every ton of paper its residents recycle, as opposed to a cost of over a hundred dollars per ton to send paper, or anything else, to a landfill. In Minneapolis, Eureka Recycling teamed up with the city to invest in local community outreach focused on keeping their recycling stream clean of contamination. Minneapolis now has one of the lowest contamination rates of any city in the country, with Eureka sharing the profits with the city. Cities around the country create significant local economic value with best-in-class recycling service, including Sims Municipal Recycling and Pratt Industries in New York City, Lakeshore Recycling Systems in Chicago, Recology in San Francisco and Seattle, Rumpke in Ohio, and Balcones Resources in Austin.

Yet it's shocking how long it took America to start building this infrastructure—and that such a large number of American communities still lack advanced recycling and other circular economy infrastructure, such as composting programs.

The burning of fossil fuels has received the lion's share of attention in the climate change debate, which has led to vital progress. The now rapidly advancing transition to solar and wind power is immensely important. But an estimated two thirds of greenhouse-gas emissions come from the linear processes of extraction and mining, manufacturing, and disposal of consumer products.

The wasteful and environmentally catastrophic linear system was developed in the twentieth century specifically to enrich companies that ginned up their profits by extracting more and more natural resources—oil for making plastic, ore for metal, and timber for paper—without being held accountable for the environmental damage they caused. They also boosted profits, at the public's great expense, by manufacturing products not for optimal longevity but with the aim that they would either soon become obsolete or be trashed after a single use. That, in turn, forced additional extraction of natural resources for each new product manufactured. As I'll reveal more fully in the first chapter, the notion that products and their packaging should be carelessly thrown away after one use rather than repaired, reused, or recycled was implanted in the public consciousness through ad campaigns. So was the allure of "trading up" to new products before they needed replacement. Unbeknownst to taxpayers, the firms responsible for this have been able to shunt these expenses off on us; many of the worst offenders, such as fossil fuel extractors, have insidiously lobbied for and gained hundreds of billions of dollars in federal subsidies. The public has unknowingly paid billions of tax dollars to subsidize the development and growth of industries that benefited from the take-make-waste economy.

There is no good reason that we should continually pay a fee for the extraction of a natural resource every time we use a product or for its disposal after we use it. We have been scammed into paying unnecessary costs for the past seventy-five years, while the land, air, and water that we collectively own has been despoiled.

The damage done to the planet, and to our societies, is becoming shockingly clear. Climate change is progressing even more rapidly than anticipated. More frequent and severe droughts are contributing to increasingly devastating forest fires. The massive conflagrations not only release huge volumes of carbon into the atmosphere, they also drastically reduce the volume of carbon the decimated forests pull out of the air and destroy the homes of hundreds of thousands of people annually. Rain forests, which are the most powerful carbon pullers, are being depleted at the rate of an estimated 31,000 square miles a year. Research shows that both the record-breaking heat wave that hit Europe in the summer of 2020 and the torrential rains of Tropical Storm Imelda, which caused severe flooding in Texas that September, were intensified by climate change. The United Nations estimates that climate-related water shortages will plague two thirds of the world's population by 2025.

For many communities all around the world, the effects have already been devastating, and they've been felt disproportionately in poorer areas and by indigenous peoples. As the Fourth National Climate Assessment, issued by the U.S. federal government, revealed, people living in poorer neighborhoods in the country experience the greatest exposure to both pollution and property damage due to extreme weather events. Toxin-emitting factories are concentrated near poor, mostly minority, neighborhoods. For example, *Fortune* reported that in the West Louisville section of Louisville, Kentucky, where 80 percent of the population is Black and the median income is $21,500, and which has the worst air quality of all midsize American cities, the air is tainted by fifty-six toxin-spewing facilities. Residents of West Louisville live on average 12.5 fewer years than do the white residents of the city's more affluent neighborhoods.

As for indigenous peoples, the United Nations reported on the wide-ranging effects of looming water shortages due to glacial melting in the Himalayas; droughts and punishing deforestation in areas of the Amazon populated by indigenous groups; the depletion of reindeer, caribou, seals, and fish Arctic peoples rely on; and sand dune expansion and drought impinging on cattle and goat farming in Africa's Kalahari Basin.

Yet even as proof of devastation has mounted, resource degradation has escalated in the past decade. A third of Earth's soil has already vanished, and if current rates of depletion continue, the planet will run out in sixty years. The rate of species extinction is accelerating, with an estimated 20 percent of land-based animals having been killed off since 1900, 40 percent of amphibian species, and another 1 million species now under serious threat of extinction. As a steady stream of horrifying photos of whales, dolphins, and sea turtles washed up onshore with their stomachs crammed full of plastic have revealed, our oceans are devastated by plastic refuse. Having discovered that plastics are breaking down into microunits, researchers have found that they have made their way to every corner of the planet, and also into our drinking water. As the chairman of the UN's Intergovernmental Science-Policy Platform on Biodiversity and Ecosystem Services said about an alarming 2020 global biodiversity assessment, "We are eroding the very foundations of our economies, livelihoods, food security, health, and quality of life worldwide."

In the face of overwhelming evidence of the damage they've wrought, many of the fossil fuel, mining, and manufacturing companies as well as most large landfill owners have fought furiously against all steps at remediation. At Recyclebank and in New York City, I had a front-row view of the underhandedness with which they've spread lies and thwarted change. I saw how progress in expanding and improving recycling and in reducing the use of environment-debasing materials has been stymied. When Mayor Bloomberg and I proposed a ban on styrofoam, for example, we were attacked with a disinformation campaign. In the midst of the COVID-19 crisis, the proplastics coalition shamelessly promoted the utterly baseless assertion that reusable bags would spread the virus, seizing on what they saw as an opportunity to overturn plastic bag bans. (The press coverage of that issue can be traced to a press release issue by the deceptively named lobbying group, the American Progressive Bag Alliance.)

Advocates of the take-make-waste system have characterized the linear economy as the optimally efficient free market. But there is nothing efficient about the fact that approximately 90 percent of plastic, ends up piling up in landfills and oceans when so much of it could be recycled. (As we'll see, many large corporations are clamoring to buy certain recycled plastics.) There is nothing efficient about the trashing of approximately forty-two pounds of electronics goods—the fastest growing part of the waste stream— per American annually, when so many of those items could be refurbished and resold. There is nothing efficient about 40 percent of food bought by Americans going to waste, a great deal of it dumped when it's still good to eat.

All of that waste, if recovered and repurposed, could be taking the place of a large portion of the precious natural resources that extractive and polluting companies are depleting to such an alarming extent. Recovering and reusing those materials, and making use of less of them in the first place, would dramatically reduce greenhouse gas emissions.

The concept that the linear economy is the epitome of capitalism was always a scam. Critics of the green movement have complained for decades that calls to make companies clean up their act are anticapitalist. They supposedly violate magical, hidden-hand, free-market mechanisms. But what's actually been hidden are violations of free-market principles, such as those enormous subsidies paid to extractors.

Gaping tax breaks have been awarded to the extractive and fossil fuel industries for decades. As Eric Schlosser wrote in his blistering *Fast Food Nation*, "legislation passed by Congress has played a far more important role in shaping the economic history of the postwar era than any free market forces." What could be less capitalistic than private companies relying on taxpayers to fund the disposal of their product, at the conclusion of one life cycle of a product's use? It's not even only the customer of a product who has been expected to pay: the system has been rigged so that every taxpayer, regardless of whether they are a product's customer, shares in the cost of its disposal. This is an anticapitalist tragedy of the commons if ever there was one.

The American Sustainable Business Council recently issued a hard-hitting missive titled "There Is No Going Back," acknowledging that "many across the country view our current capitalist system as rigged and not working for them." Rightfully so, and that's true in far too many economies around the globe. It's time to dismantle the rigging. It's time to transition to a circular economy. The viability of life on the planet depends on it.

The Urgent Circular Solution

What is a circular economy? It's an economy that invests in advanced technologies related to material science, product design, recycling, and manufacturing that leads to a zero-waste "closed-loop" system in which resources are not wasted.

Products are manufactured with locally generated, renewable, clean energy, and made from sustainable materials, repurposed and recycled materials, or a wealth of nonpolluting biodegradable materials that are being developed. They are designed for longevity—not planned obsolescence—and for reuse and repair. Money is no longer wasted and lives are no longer harmed by relying on extracting natural resources and disposing of products in landfills. Design and manufacturing are optimized to be in harmony with nature and with the people who use them and the communities they live in.

Transitioning to circular methods of production, distribution, consumption, and reuse of products and materials will heal our planet and generate enormous economic opportunities. Innovative entrepreneurs can build transformative and fast-growing businesses with circular solutions. Corporate behemoths can reduce materials acquisition, packaging, and transport costs, while delighting customers with their planet-friendly progressivism. We consumers can save considerably by purchasing longer-lived products, getting them repaired and upgraded, and purchasing only the service we want from a product, rather than paying the excessive expense of actually owning the product. A circular economy ensures that consumers, consumer goods companies, and municipalities are owners of an economy that works to their benefit. It does away with the hundreds of billions of dollars in costly extraction

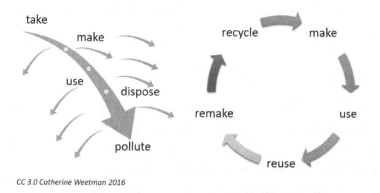

CC 3.0 Catherine Weetman 2016

Linear Economy vs. Circular Economy

and landfill fees that we have been paying for the past seventy-five years. Equally important, it prohibits the theft of our oceans and our land by polluters who claim it's necessary for the development and growth of our economy. It is not. In a circular economy, scamming is not rewarded, but merit, hard work, transparency, and innovation are.

While the transition to circularity might seem improbable, due to entrenched corporate interest in taking and wasting, the business case for circularity is in fact the most powerful driver of its increasingly rapid adoption. As important as government action is, it's business innovation that will lead to transformative products and services—hand in hand, that is, with consumer demand for them.

Companies are now highly motivated to transition to circularity. That's not only because the brands that have led the way have seen fantastic improvements in profits and consumer loyalty, but also because of the accelerating effects of climate change. More and more companies are facing supply-chain challenges as the depletion of natural resources drives up production costs. They're also buffeted by increasingly frequent natural disasters, from flooding to water scarcity, fires to seawater rise. Brands are also increasingly aware that the public will hold them accountable for their environmental devastation and rearguard action against green solutions.

The Distinctive Superiority of Circularity

In my role now as founder and CEO of Closed Loop Partners, the first investment firm dedicated entirely to financing the adoption of circular solutions, I'm constantly asked whether "circular economy" isn't really just the latest buzzword for sustainability. No is the answer. Sustainability innovations are making vital contributions to the development of circular economy products and business models, but the concept of a circular economy is distinctive, aiming not only for sustainability but for renewal. Its origins date back to the 1960s, but it's only recently come to prominence among corporate business leaders, financial investors, and government and NGO figures, moved forward by a legion of entrepreneurs devising circular products and services. That's because they've recognized that circularity is superior for profits while also potentially transformative in battling climate change.

The case for the superiority of a circular, closed-loop system has been advanced with particular vigor by one of the founding thought leaders of the circular economy, the Swiss architect Walter Stahel.

He founded the Product-Life Institute to advance practices of circular production, where he's consulted with businesses, government, and NGO leaders for decades. He was also instrumental in articulating the core set of operating principles for circular production:

Reduce—Continually cut down on the amount of natural resources used, the waste generated, the environmental damage done, and the amount of greenhouse gases emitted.

Reuse—Build for durability so products and their packaging can recirculate to new users, with little or no refurbishment.

Remake—Repair, refurbish, and more substantially remanufacture.

Recover—Design products for easy disassembly and repurposing of materials, and develop "reverse logistics" by which manufacturers and retailers take their products back for either refurbishment and resale or recycling.

Renew—Use only renewable energy, work with regenerative methods of production, and construct the built environment so that it actually replenishes resources.

As we'll explore, a number of other streams of thought have contributed to the development of the circular economy concept and have provided tools for innovating solutions.

The Movement Comes of Age

I was thrilled to learn about the rich and wide-ranging innovations circular economy proponents have advanced, and when I was working for New York City I was intent on implementing as many circular solutions as I could. My first initiative was a curbside food waste composting program, which I was gratified to find New Yorkers embraced. As I searched for solutions and collaborators, I learned about a tidal wave of entrepreneurship advancing circular innovations, which I'll profile in the chapters ahead. They include the invention of many new biodegradable materials, like packaging made out of mushroom roots, to replace foam. Another is fabric made from algae, grown in a zero-carbon-emissions, solar-powered, and extremely water-efficient system. The fabric is not only biodegradable; it also releases valuable nutrients into your skin, like fabric body lotion.

Breakthrough advances are being innovated not only by start-ups, but by many of the largest enterprises on the planet. They include consumer goods behemoths Unilever, Procter & Gamble, Nestle, Coca-Cola, and Pepsi; IKEA in furnishing; Google, Dell, and HP in computing; carmakers Ford, GM, and Renault; food giants Kroger, Starbucks, and McDonald's; and even a pioneer of the wasteful fast-fashion revolution, H&M.

Take the case of Unilever, which is one of the world's largest users of plastic packaging. Among many other initiatives, it's made the following four commitments regarding plastic:

- Reduce virgin plastic packaging by 50 percent by 2025, with a third coming from an absolute plastic reduction.
- Help collect and process more plastic packaging than we sell by 2025.
- Ensure that 100 percent of plastic packaging is designed to be fully reusable, recyclable, or compostable.
- Increase the use of postconsumer recycled plastic content in packaging to at least 25 percent.

These leading brands are excited to contribute to healing the planet, but are also spurred by the business opportunities of circularity. They understand that consumers are deeply serious about putting their money into purchasing planet-healthy products—especially true of millennials. They now comprise over half the work force and wield as much purchasing power as baby boomers, and 75 percent of them are already favoring the purchase of sustainable goods. Unilever's success with its green products exemplifies how strong consumer demand is. In 2018, Unilever's Sustainable Living Brands grew 69 percent faster than the rest of the business, up from 46 percent faster in 2017.

Even in extractive industries, some major commitments to circularity have been made. A pioneer was India-based Tata Group, one the world's largest companies, which among many other enterprises is a leading producer of steel. In 2001, Tata Steel and the Steel Authority of India established the joint venture mjunction as an e-marketplace for by-products (e.g., secondary steel) and idle business assets. What would be waste is now feedstock for companies, and mjunction has become the world's largest e-market for steel, making it India's biggest B2B e-commerce company.

Results like those have led to a surge of awareness of the huge benefits of transitioning to circularity in the global business community. For example, the World Economic Forum made the circular economy the centerpiece topic of its 2019 Davos summit, and leading global business consultancies McKinsey, Deloitte, and Accenture are all promoting circular economy innovation and providing services for making the transition. The scale of the economic opportunity is staggering. Accenture calculates that the transition to a circular economy around the globe that could be achieved between now and 2030 would lead to $4.5 trillion of economic growth, describing it as "the biggest opportunity to transform production and consumption since the First Industrial Revolution 250 years ago."

One problem, though, is that there is still so much confusion about what the transition to circularity entails. For example, investment bank ING conducted a survey of business leaders in the U.S. that showed that 62 percent of them want to adopt circular processes, but only 16 percent of companies have actually begun to take any significant steps. Meanwhile, those efforts are mostly limited to cutting costs on materials rather than innovating new circular products and services. That need for clarity is my main motivation for writing this book, and I've devoted a chapter to telling more of the story of how the various threads of research and argument have come together to form the full-bodied contemporary understanding of what a circular economy entails.

Another reason the transition hasn't progressed further yet is that more financing is needed to support innovators. That's why I started Closed Loop Partners. I saw that there were so many great solutions being developed that only needed a shot in the arm of financial support in order to gain traction and scale up. I decided that Closed Loop Partners would focus on funding those that my team and I identified as having the most transformative potential. I'll introduce a number of them and their ingenious founders throughout the book, as well as many other bold thinkers and business builders leading the way forward.

We've launched a number of innovation initiatives in partnership with major brands. They include our NextGen Consortium, to support the wide-scale commercialization of planet-friendly replacements for nonrecyclable paper cups, with McDonald's and Starbucks; and our Consortium to Reinvent the Retail Bag to find the optimal replacement for the single-use plastic bag menace, with partners including Walmart, Target, Kroger, Walgreens, and CVS Health.

This activity signals that we've reached a tipping point in advancing circularity. But powerful counterforces are still vigorously working against the implementation of policies to speed change, as well as against the raising of public awareness to fuel demand for it. Self-serving corporate players with vested interests, and governmental and media abettors, use stealthy tactics to delay transformation, regrettably to powerful effect. A rearguard lobbying campaign has stalled progress, for example, in passing a number of right-to-repair laws put forward in the U.S., which would force manufacturers to make products like mobile phones much more easily and cheaply repairable. These lobbyists include some of the major medical equipment companies, which, as we'll see, restricted repair information for ventilators and other life-sustaining equipment as the COVID-19 pandemic spiked.

Probably the most pernicious feat the take-make-waste defenders have achieved, though, was to canonize in boardrooms, universities, and newspapers the deeply flawed notion that ever-increasing consumption is the necessary engine of national economic strength and the fundamental source of individual well-being. As I'll chronicle, convincing Americans, and then global consumers, to go along with this "gospel of consumption" was no easy task, as it goes against the grain of a long-standing respect for and practice of thrift. In the first part of the book, I tell the story of the decades-long campaign to convince Americans that throwing things away is a life-enhancing convenience, even a patriotic duty, conducted by a cohort of brilliant persuasion experts. I then pull back the curtain on crafty public "service" campaigns and insidious disinformation lobbying that kept the public largely in the dark about the environmental devastation of the take-make-waste system.

As recent revelations show, the ozone-evaporating effects of fossil fuel burning and the ocean-contaminating infiltration of plastic refuse were understood by the fossil fuel behemoths and plastics producers more than fifty years ago. As I'll also chronicle, they produced bogus research and promoted diversionary "greenwashing" solutions that duped the public. I'll draw on my personal experience working for New York City to portray how underhanded, and menacing, these special interest operators can be.

Understanding disinformation is vital to us all becoming more energized and effective advocates for change. Learning more about these tactics as I've worked on this book has certainly opened my eyes even more. But the most powerful means that we in the public have to accelerate the advance of circularity is through our purchases, and the good news is that so many wonderful products and services are rapidly coming to market. In the second part of the book, I'll take you on a journey to meet many of the most inventive and influential innovators.

The wealth of innovation going on should give us all great optimism about the progress to come. As Peter Diamandis and Steven Kotler wrote in their wonderful book *Abundance: The Future Is Better Than You Think*, "the greatest tool we have for tackling our grand challenges is the passionate and dedicated human mind." The many brilliant minds you'll meet in the following pages are a powerful testament that we have the tools we need.

PART
ONE

Defeating
Take
and
Waste

A Duty to Waste

When ER nurses are scared, THAT is when I am scared. I am an ER nurse, and I absolutely, without a doubt, am truly and completely frightened.

I am scared because two weeks ago, it was absolutely frowned upon not to immediately throw away gown and mask after coming out of a contact/droplet patient's room, and this week "wear your same mask all day" is the new norm.

S O WROTE ONE NURSE in one of many pleas nurses sent to newspapers as cases of COVID-19 spiked dramatically in the U.S. in April 2020, unnecessarily and tragically causing tens of thousands of deaths. A coalition of doctors spontaneously formed the organization Get Us PPE, working desperately to salvage supplies. When they sent a survey to health-care providers asking "What supplies do you need?" representative responses from the over seven thousand nurses who answered were "We are out of everything," and "Providers using one mask for 3+ weeks."

The shortage of personal protection equipment (PPE) was entirely a result of the take-make-waste economy. PPE was designed to be trashed after one use, and no systems were implemented for sterilizing it for reuse. There was, it turned out, no good reason for that; sterilizing with a mist of hydrogen peroxide was one simple and inexpensive method quickly devised. Hospitals simply hadn't questioned this inefficient use of funds, despite many being strapped for cash. Throwing away PPE was a legacy of a horribly inefficient business model, not a medical requirement. The result was that doctors, nurses, and first responders unnecessarily lost their lives.

The PPE shortage was just one of many appalling failures of the world's largest economy to meet the demands of the pandemic. The coronavirus outbreak exposed numerous flaws in supply chains, with farmers dumping tons of vegetables and milk because the complex supply chains they sold into had broken. Meanwhile, people in the nearby communities scavenged almost-empty grocery store shelves. Appalling injustices of economic inequality were also laid bare. An estimated 10.5 percent of American households suffered from food insecurity in 2019, according to the U.S. Department of Agriculture. Many children are able to eat a healthy breakfast and lunch only because they're served them at school. Suddenly, with schools closed, they are facing a food crisis.

As I watched horrible inadequacies come to light, I thought of an article, "The Capitalist Threat," written by one the world's most successful investors and leading philanthropists, George Soros. I read it while in college in the 1990s, and it had a profound effect on me. Soros offered a lacerating critique of "the untrammeled intensification" of dangerous market forces that have perverted our system, and how, ultimately, they may imperil the foundation of our democracy. The so-called perfect knowledge of the market, Soros wrote, wasn't actually serving the public's authentic needs and desires. "Advertising, marketing, and even packaging, aim at shaping people's preferences," he noted, "rather than, as laissez-faire theory holds, merely responding to them." I was studying history and economics at the time, becoming increasingly incredulous about all the theories of market perfection and meritocratic economic opportunity. The picture of the economy my professors painted looked nothing like the reality I'd seen growing up in a tough neighborhood of Philadelphia—the best my mother could afford after my parents divorced. She worked hard as a teacher, underpaid and putting in many extra hours on nights and weekends; but it was clear she was never going to get off that grueling treadmill. The American meritocracy that had once afforded massive opportunity had been corrupted. It was empowering for a young college student to read one of the greatest investors articulate that realization in a clear thesis calling for an upgrade to how our economy functioned.

How had our economy, which purported to prize hard work and free market forces, become so distorted that the self-proclaimed champions of capitalism were, in fact, often secretly collecting huge government subsidies and blocking competition in order to protect their outdated business models and profits? How had we been convinced that such a broken, unbalanced system was serving our interests?

Thrifty Heritage

When the home of John Thoman in Woodside, New York, was raided and four filled containers were discovered, he was lucky, let off with only a warning. Joe Wittman, in nearby Woodhaven, was not so fortunate when two weeks later another raid uncovered twenty-three filled containers and 263 empty ones in his home. He avoided arrest but received a hefty fine. Poor thirteen-year-old Carmen Carmenello was arrested for selling four empty containers, and was remanded to the Children's Society.

The year was 1916, and these highly prized commodities were glass soda, beer, and mineral water bottles, ownership of which was fiercely protected by the Long Island Bottler Association. As the association crowed to its members in its annual raid report for that year, the Detective Department was vigilant in serving their interests—conducting a total of forty-three raids, a rate of close to one a week. One can only hope they were investigating burglaries and murders with comparable vigor.

When I read the report, I was struck that used glass bottles were once seen as so valuable that people were actually willing to risk arrest to hoard and reuse them. Today, a vast quantity of glass bottles are thrown away in landfills. New York City had seriously considered ending its glass recycling a few years before I was hired, because, for reasons I'll dive into later, it was so hard to find a good market for the used glass. But at the turn of the twentieth century, bottlers wanted every single bottle back for refilling; and they came close, with a 95 percent return rate. Those who didn't hoard bottles happily dropped their empties off at their grocers, strongly incentivized by the 1- or 2-cent deposit they could collect for each, which may not sound like much but accounted for 40 percent of the cost of the drink; the equivalent of about fifty cents in today's money versus our 5-cent deposit. Today, we spend taxpayer dollars to send about two-thirds of the glass bottles used every year to landfills.

Such profligate tossing is second nature to us now, but it's an aberration in the long sweep of human history. For eons we practiced circular production, and we were extraordinarily resourceful about it. The concept of paying for packaging every time one buys a product would have seemed like a scam. Historian of early human invention Maikel Kuijpers writes that our earliest forebears in the Paleolithic era, which began 3.3 million years ago, repurposed every tool they made, for instance by breaking down worn ax blades into small flint tools. In the Neolithic age that followed, when our ancestors began making pottery, despite how plentiful the clay was they still ground broken pots into powder for making new ones. Even stones would be repurposed, perhaps used first as grindstones, then as doorsteps or tombstones. After metalworking was invented in the Bronze Age, metal objects would be melted down and recrafted innumerable times. The biblical commendation to beat swords into plowshares was no matter of mere poetry.

One of the great leaps in human progress, the invention of paper in 105 AD, was accomplished by recycling. The minister of agriculture in China's Han dynasty figured out how to dissolve old linen rags into a pulp, which he spread into a film and dried to make thin paper sheets, providing rag collectors with gainful employment for centuries thereafter. During the Middle Ages, as grand cathedrals were built, sculptors recarved statues of Roman gods into figures of Christian saints. Even when the European courts became centers of lavish wealth, courtesans sent their gowns and ornamented waistcoats back to tailors to be refashioned for the next season. All the way into the middle part of the twentieth century, the reuse and repurposing of goods and materials of all kinds was a normal part of everyday life. Robust systems were in place for collecting, repairing, and reselling almost everything. Peddlers that sold new products to households also bought repairable items in order to sell them. As more manufactured products of all sorts— from cameras to gas stoves and tractors—hit the market when the Industrial Revolution heated up, repair specialists made good livings breathing new life into manufactured goods. Shoes were repeatedly resoled, razor blades were resharpened, and pens were refilled.

So when and why did we begin to prefer throwing so much away? Who convinced us that masses of totally unnecessary waste are the natural consequence of living a good life? How did we fall for the notion that we should pay the cost for a new package made from virgin material every time we buy a product? The task wasn't easy.

Engineering Consumers

Philosopher Marshall McLuhan, who famously coined the phrase "the medium is the message," wrote that "Historians and archeologists will one day discover that the ads of our time are the richest and most faithful daily reflections that any society ever made of its entire range of activities." Ads have, as George Soros stressed, not only reflected our behavior, they have enormously influenced it. In the words of one of the innovators of modern advertising, Earnest Elmo Calkins, "Wearing things out does not produce prosperity. Buying things does." Calkins was one of the master-minds of consumerism, or, as he dubbed it in a 1930 paper explaining the science of artificially stoking demand for products people didn't need, "consumer engineering."

He and his fellow admen cooked up the notion of attaching the word "consumption" to nonfood products in the 1920s. Before then, it had only been used to refer to food. In a masterstroke, Calkins and crew managed to popularize the ironic notion of consuming durable goods, like stoves, refrigerators, and radios, instilling the cultural idea that just like food, all products have a shelf life. Calkins explained the strategy thus: "Goods fall into two classes, those that we use, such as motorcars or safety razors, and those that we use *up*, such as toothpaste or soda biscuits. Consumer engineering must see to it that we use *up* the kind of goods we now merely use." And so it was that we citizens became "consumers."

The notion that all manner of products should be disposed of, rather than reused and repaired, had actually been gaining ground for a couple of decades by then, beginning with the introduction of packaging that was specifically designed to be tossed. The restaurant chain Cracker Barrel gets its name from the barrels grocers sold loose crackers in, to be scooped up by customers into their own bags and bins. Then, in 1899, as Susan Strasser highlights in her history book *Waste and Want*, the market-leading U.S. cracker maker, National Biscuit—soon to become Nabisco—introduced its patented In-Er-Seal packaging, wrapping a new brand of much lighter and flakier crackers, Uneedas, in waxed paper and shipping them in cardboard boxes. If they were sold in the big barrels, they would quickly become soggy. Shipping them to stores in the newfangled hermetically sealed slips kept them crisp. National

Biscuit spent a fortune advertising their new marvel, launching a $7 million campaign that brilliantly hawked the waterproofing by featuring a young boy in a yellow rain slicker carrying a box of Uneedas through slanting rain.

Next came throwaway tin and steel cans, relieving women of the arduous job of preserving fresh produce in glass jars for their families. Though food in metal cans had been introduced commercially in the 1860s, it had been much too expensive for all but the very wealthy and was primarily sold to feed armies. However, in 1904, the Max Ams Machine Company invented a method for mass-producing tin cans so inexpensively that wide-scale commercialization began.

Unlike glass bottles and canning jars, which were made for reuse, processed food cans, with their metal lids, were specifically designed to be opened only once, then chucked. The American Can Company, which quickly cornered 90 percent of the market, made no attempt to reclaim them. Fast-forward to today, when packaging and containers account for 32.5 percent of municipal waste in the U.S.

Convincing the public of the benefits of throwing products away proved a trickier challenge than the new consumer engineers expected. Brands started small. Disposable rubber gloves, promoted first for use in surgery and only later for household use, were invented in 1894. Gillette introduced the first disposable razor blade in 1895. A machine for making paper plates efficiently was invented in 1904, but the idea of throwaway plates was originally perceived as ridiculous. It took decades for them to be widely purchased by the public.

Most Americans were still accustomed to durability being a core characteristic of a product, but the campaign to convince Americans that consuming and disposing produced a better lifestyle began to gain traction during World War I. While the economies of the European combatants were wracked by the war, the American economy boomed. In 1914, at the war's start, the U.S.

was in the midst of a deep recession. Massive orders from Europe of both agricultural and manufactured goods jump-started an astonishing economic turnaround. Exports to Europe rose from $1.479 billion in 1913 to $4.062 billion in 1917. After the country entered the fight that year, federal spending on war matériel spiked from $477 million the previous year to $8.5 billion by 1918. Unemployment fell off a cliff, from a staggering 16.4 percent in 1914 to 1.4 percent at the end of the war, and average weekly earnings doubled. While the government urged thrift, promoting ideas that families take part in "meatless Mondays" and go back to canning, and that earnings windfalls be poured into Liberty Bonds to fund the war, some retailers kept up a steady stream of ads suggesting that the public had a duty to spend. Shops hung signs in their windows saying BEWARE OF THRIFT AND UNWISE ECONOMY and CLEAR THE TRACK FOR PROSPERITY! BUY WHAT YOU NEED NOW!

Advertisers blanketed papers with ads portraying the purchase of products as a benefit to the war effort. Sometimes the pitch was quite a stretch. One was an ad for Nemo corsets, which proclaimed that "Women who work, especially those who are doing unaccustomed war-time labor, must guard their health to retain their efficiency," which meant that the corsets were "now, even more than ever—a national necessity!" In the words of advertising guru of the time Frank Presbrey, the war led to a "new and greater revelation of the power that advertising possesses."

With war's end, that power was trained like a laser on convincing Americans to throw away their stodgy old pots and pans, their antiquated irons and washboards, their time-consuming brooms and dustbins and buy the bevy of new mass-produced electronic household products that streamed out of America's war-enhanced factories. Having learned very well from the war just how lucrative mass production could be, manufacturers repurposed their assembly lines for producing "consumer goods" with a fury. The miracle of mass production required mass consumption, and marketers convinced Americans they should spend profligately. Spurring unnecessary consumption and making products for disposability rather than durability became the lynchpins of success.

Planning Obsolescence

When most producers took great pride in building things to last, many did a marvelous job of it. Take the case of one electronic product that, in time, we came to expect would quickly go kaput: incandescent lightbulbs. One bulb from the first years of production is still in operation. Dubbed the Centennial Light, it was first switched on in 1901 in a fire station in Livermore, California. The station has installed a webcam and streams video 24/7 of the bulb dimly shining, and the station chief reports that the bulb has now outlasted three webcams. So why did lightbulbs become so short-lived?

In 1924, a group of executives from the leading lightbulb manufacturers, including General Electric and Philips, journeyed to the luxurious Swiss enclave of Geneva and formed what came to be known as the Phoebus cartel. The companies created a laboratory in Switzerland and combined forces to develop a new standardized bulb that would burn for no more than a thousand hours and would break more easily—requiring the sale of more bulbs. Members agreed to a strict set of "degradation guidelines" and were required to regularly send samples of their bulbs to the Swiss lab for testing. If they were found to be too long-lived, the firm was slapped with a considerable fine. Some members did try to game the system, selling longer-lasting bulbs to burnish their reputations, which the CEO of Philips* bemoaned in a letter to a GE coconspirator: "After the very strenuous efforts we made to emerge from a period of long life lamps [as bulbs were called then], it is of the greatest importance that we do not sink back into the same mire" of "lamps that will have a very prolonged life." This type of

---

*As an example of redemption, new leadership, and forward planning, Phillips is now recognized as one of the leaders in circular economy business models.

thinking among CEOs is neither capitalist nor socialist. It's simply thinking about the best way to scam consumers.

The Centennial Light, by stark contrast, was clearly built for longevity. Why aim for long life? Because lightbulbs were at first owned by the electric companies, which installed lighting systems in homes and retained ownership of all the fixtures, replacing the parts, including bulbs, at their own expense. But as electricity went mass market, the companies learned they could make more money by selling fixtures, lamps, and bulbs to their customers, and the more sales the better. This so-called repetitive sales model became the driving force of economic growth in the Roaring Twenties, and was taken to mind-boggling extremes in the decades to follow.

Ironically, some of the best evidence that the practice became widespread comes from the indignant responses of product designers to an article forthrightly titled "Product Death Dates—A Desirable Concept?" published in a leading journal for product engineers, *Design News*, in 1958. Written by the journal's editor, the article asked, "Is purposeful design for product failure unethical?" and cited an engineer from a radio manufacturer, who proudly told him the firm's radios were designed not to last more than three years. "Should engineers resist such a philosophy," the article continues, "if their management specified that it wanted a 'short-term product?'" concluding that no, they should not. Responses flooded in from product engineers. One, from Fairchild, then a maker of cameras and later a semiconducter pioneer, attempted to defend his fellow engineers by objecting that planned obsolescence was "practiced by nearly all design groups, in all fields, under the guise of economy or efficiency." He went on to argue that "it is wasteful to make any component more durable than the weakest link, and ideally a product should fall apart all at once." His course of logic would mean that if a person is driving down the highway and their taillight burns out, the whole car should fall apart. Why not simply design for all short-lived components to be easily replaced—as car taillights are?

The making of products too difficult or costly to repair may well have been the most winning means of imposing the equivalent of death dates. It made the purchase of a new model the much preferred, and obviously rational, choice. Apple has taken this tactic to a new level, claiming that repairing its smartphones is actually dangerous—opening up a phone might cause the battery to burst into flame. If that is the case, surely with all of the amazing innovations added to the next generation iPhone every year, shouldn't a safer battery be one of them? If obsolescence hasn't been a goal, why was Apple the first to make a battery-powered device whose battery couldn't be replaced—the iPod? Only two years after I bought an iPhone 4, I discovered Apple had discontinued production of its charger. Obsolescence can be imposed in many ways.

But the truly clever practitioners of obsolescence figured out that they didn't have to risk their customers' ire; they needn't go to the trouble of designing for disuse. They could impose obsolescence by desire, making customers want new versions of products so intensely that they couldn't wait to trash their current ones. The mastermind of this "dynamic obsolescence," as he called it, was longtime CEO of General Motors, Alfred P. Sloan. His motivation came from the overwhelming competition he faced from Henry Ford, who boasted that his Model T was so sturdy it would last his customers a lifetime. The car was no beauty, but it was, as Ford promised, built with "the simplest designs that modern engineering can devise," which gave it what auto historian Lindsay Brooke praises as its "stone-simple serviceability." The car could be repaired, he reports, "with a few simple hand tools, some bailing wire, and the most basic mechanical skill." So popular was the Model T that it is still on the list of the top ten bestselling cars of all time. Yet Sloan's strategy did it in.

In 1925, Sloan instituted annual model changes to GM cars that, as one auto historian notes, "created the illusion of technological progress . . . while leaving the mechanical realities largely unchanged." Within two years, GM had overtaken Ford in sales.

So devastated was Ford's market share by 1927 that Henry Ford was convinced he had no choice but to retire his beloved Model T and introduce the souped-up Model A.

The case had been made, and manufacturers leapt to the cause, hiring artists to craft exuberantly gorgeous coffeepots, toasters, phonographs, vacuum cleaners, refrigerators, radios, lamps, and, of course, tail fins. The field of industrial design was born, and its practitioners were enormously well paid, earning salaries of $50,000 (the equivalent of $680,000 today) far more than the average pay for a corporate executive at the time.

Lehman Brothers partner Paul Mazur, who specialized in analyzing the retail trade, summed up the triumph with what might seem sarcastic derision, but was in fact praise, in his 1928 book, *American Prosperity: Its Causes and Consequences*: "Wear alone made replacement too slow for the needs of American Industry. And so the high-priests of business elected a new god. . . . Obsolescence was made supreme. . . . It could be created almost as fast as the turn of the calendar, certainly as rapidly as the creative power of inventive minds determined."

Manufacturers were greatly abetted in the feat by admen and pioneers of the deceptively named new field of public relations [PR]. The goal of their persuasion campaigns was clearly stated by Mazur in the *Harvard Business Review*: "We must shift America from a needs-culture to a desires-culture. People must be trained to desire, to want new things, even before the old have been entirely consumed. [. . .] Man's desires must overshadow his needs."

A great deal of credit for achieving the sea change has gone to public persuasion guru Edward Bernays, often called the father of PR. He was one of the most innovative and effective manipulators, devising means of applying insights from psychology that were so successful they're still very much in use today.

Bernays's ambition was sweeping and rather sinister—to control public opinion en masse. He apparently saw his services as much for the greater good, writing that the "manipulation of the organized habits and opinions of the masses is an important element in democratic society." The success of mass production and mass consumption, in addition to that of democracy, he believed, relied on creating a mass mind. And the means of doing so were now at hand. In his treatise "The Engineering of Consent," he wrote "the tremendous expansion of communications in the U.S. has given this Nation the world's most penetrating and effective apparatus for the transmission of ideas. Every resident is constantly exposed to the impact of our vast network of communications which reach every corner of the country, no matter how remote." The first radio station had commenced broadcasting in 1920, and within just a few years, radios graced the living rooms of the vast majority of American homes.

Products should be sold, Bernays argued, not based on the superiority of their features, but on the promise that they would boost customers' happiness, enhancing their health, self-esteem, and sex appeal. After all, he asserted, "The group mind does not *think*. . . . In place of thoughts it has impulses, habits, and emotions." His claims to public service include a highly effective Lucky Strike campaign called "Torches of Freedom," designed to get women smoking cigarettes, then considered crude. Hiring a group of fashionably dressed women to march down New York's Fifth Avenue in a faux protest of inequality while conspicuously smoking, he ginned up major media attention. Another great success was his campaign to increase bacon consumption as part of a "hearty" American breakfast. In a lunatic twist of irony, he was the first to invoke doctors as product advocates, so it's to him we owe the once ubiquitous "four out of five doctors surveyed" sleight of hand. Bernays also pioneered endorsements by celebrities, promising glamour by association. For such grand national service, Bernays was commended by President Hoover for "helping to create a limitless future of American consumption."

Another particularly potent voice promoting the duty to consume was pioneering "home economist" Christine Frederick, the Martha Stewart of her day. A consulting editor of *Ladies' Home Journal*, she avidly championed throwing away the barely used to make way for the shiny new. She also praised the spanking new inventions of consumer credit and installment payment plans in her introduction to the 1925 book *Midas Gold: A Study of Family Income, "Overselling" and Time-Payment as a Broadener of the Market.* Credit cards and installment plans surely did broaden markets—they were so popular that average household debt, which had theretofore been almost entirely mortgage debt, doubled during the decade.

Midas Gold, along with Frederick's own magnum opus, *Selling Mrs. Consumer,* were advice books for businesses, not consumers, extolling the virtues of what her husband, George Frederick, the president of a business books publishing house, called "progressive obsolescence." "What is 'progressive obsolescence?'" Christine wrote glowingly. "A readiness to 'scrap' or lay aside an article before its natural lifetime . . . a willingness to apply a very large share of one's income, even if it pinches savings, to the acquisition of the new goods." She introduced to the business canon the audacious new concept that being wasteful could actually be creative, writing, "It is now time to assert and proclaim for the American family . . . a bold new policy. . . . This [is] the policy of creative waste." After all, she asserted, "There isn't the slightest reason in the world why materials which are inexhaustibly replenishable should not be creatively 'wasted.'" What she entirely failed to appreciate was that in fact those materials are not inexhaustible. But she was right that "Mrs. Consumer has billions to spend," and "She is having a gorgeous time spending it."

The horrible irony of the boom in consumerism was that even the cleverest techniques of "overselling" couldn't stoke sales enough to keep up with the zeal with which manufacturers overproduced. The buildup of excess inventories is now widely credited as a major cause of the Great Depression. *Selling Mrs. Consumer* was published in 1929, just weeks before the October 29 stock market crash. In the book, Frederick wrote "If the credit of the United States is the most solid credit in the world today, it must be because consumers make it so." The depth of the Depression that immediately followed made it clear that household debt would not disappear simply because consumers wished to make it so. Consumers also couldn't conjure up jobs in order to have the money in their pockets to consume more. The Depression was the first hard lesson of the harsh and lasting consequences of a society that values quantity over quality.

No exhortations to keep spending or all the massive stimulus programs of the New Deal could pull the country out of the downturn. Only the most astonishing military buildup in history could achieve that.

A Post–World War II Spending Frenzy

While racing toward the industrial hub of Bremen in the north of Germany, on the clear spring day of May 22, 1944, a formation of sixteen enormously powerful but highly nimble new American P-47 Thunderbolt fighter jets, led by Lieutenant-Colonel Francis Gabreski, spotted a group of German Focke-Wulf fighters just taking off from their base. With a thrill of confidence, Gabreski led his squad in rapid-fire pursuit.

Though the German formation also comprised about sixteen planes, the duel was no contest. Within moments, Gabreski shot down one Focke-Wulf and immediately turned to pursue another, whose pilot bailed rather than even attempt to return fire against the P-47's mighty guns. Gabreski and his men shot down thirteen Focke-Wulfs in the engagement and damaged three more, while suffering only two losses.

The squad's dogfight that day is exemplary of the crushing superiority of the U.S. war machine by that time, the product of an industrial expansion that was truly awe-inspiring. American planes were pouring out of factories at a mind-boggling clip. The massive Ford Willow Run facility, built in 1942, had created such an efficient assembly line by 1944 that it was producing a B-24 bomber—with 1.2 million parts to be meticulously pieced together—in just one hour. Most factories were operating twenty-four hours a day, seven days a week.

Upon U.S. entry into the war, President Roosevelt had told the nation, "It is not enough to turn out just a few more planes. . . . We must outproduce them overwhelmingly." American factories most definitely did. Luftwaffe commander General Adolf Galland reported to his superiors in April 1944 that "the ratio in which we fight today is about one to seven." Before the war, the U.S. military boasted just 72 fighter jets, and Roosevelt had shocked the nation by proclaiming 50,000 aircraft would be needed. By war's end, production stood at 297,000. And that was only the planes. U.S. munitions makers also produced 806,073 military trucks; 86,338 tanks; 76,400 ships, from entirely new forms of landing craft to the first aircraft carriers; 17,400,000 firearms; and 41,400,000,000 rounds of ammunition.

In order achieve this massive buildup, the manufacturers needed continual access to precious raw materials. In response, the federal government launched a major advertising campaign to promote recycling and encourage families to grow their own veggies in Victory gardens, as well as a return to home canning. Ubiquitous ads and posters prompted SAVE YOUR CANS, FOOD IS A WEAPON, DON'T WASTE IT! Scrap drives exhorting housewives to donate pots and pans and kitchen utensils for smelting brought in a haul estimated at six million kitchen items.

Looking back, we can reflect wistfully on the prospect that after a decade of the Great Depression, and considering the critical role repairing, reusing, and recycling had played in helping the U.S. win the war, gonzo American consumerism had seen its zenith in the 1920s. But the ideology that consumption was the route to prosperity was to be reintroduced, even before war's end, with devastating consequences for household debt as well as our environment.

As the U.S. emerged after the war as the global economic leader, incomes boomed. Seventeen million new jobs were created, and the national average weekly salary increased between 50 and 65 percent. The preachers of the gospel of continuous spend and consume were poised for an unprecedented spending spree. Americans had contributed $185 billion in bonds to the war effort, and those bonds would be coming due, creating a massive cash windfall. Manufacturers had prepped the public to start spending with abandon. The take-make-waste model was about to make a huge comeback.

Many firms ran ads toward the end of the war promoting soon-to-come new wonders, such as home movie cameras and projectors. The 1943 "Kitchen of Tomorrow" campaign featured the Therm-X oven, specially designed to heat "ready-to-eat" packaged meals. The campaign caused quite a stir, inspiring feature stories in *Life* and *Better Homes & Gardens*, a Paramount Pictures short film, and a traveling display tour that was visited by 1.6 million people. Buying kitchen goods was becoming a form of entertainment.

Then there was the new allure of television. Televisions had gone on the market in 1939, but production was forbidden during the war. While by 1942 only five thousand had been sold, after just one year of production following the war, forty-four thousand had entered American homes, and by 1948 that rose to 2 million. They were not only hot commodities, of course; they fast became the hot new advertising medium. By 1951, television ad revenue totaled $41 million, and in two years it had grown eightfold, to $336 million.

The selection of disposable products boomed, many made of cheap new plastics. *Newsweek* insightfully predicted in 1943 that we'd live in a "plastic postwar world." Why wash silverware? Buy plastic utensils and throw them away! Why fuss with refilling a pen; get plastic ones and just toss them. Here came plastic cups, toothbrushes, razors, sandwich bags, straws, and bottles. While the makers of glass bottles had been so intent on reclaiming them that they'd had people arrested, those who first produced plastic bottles made no pretense they should be returned or recycled.

Glass bottles too became disposables postwar, and stamped "No deposit, no return."

Aluminum, infinitely recyclable into new products and zealously collected during the war for recycling, was molded into disposable frozen-food containers, popcorn popping bags, toss-away grills, frying pans, and even disposable dog dishes. Paper plates finally took off, along with paper napkins, tablecloths, and towels. A 1955 *Life* magazine article titled "Throwaway Living" featured a photo of a gleeful couple tossing disposable household items in the air and claimed that, combined, the items in the picture would save forty hours of washing-up time.

As historian Sheldon Garon explains in his book *Beyond Our Means*, the federal government helped supercharge consumption by loosening controls on installment plan purchases, encouraging buying on credit by making the interest charged tax deductible. Government economists actually characterized installment buying as a form of saving. Delighted by the advent of the baby boom, the Commerce Department installed a display in its lobby, Garon reports, that celebrated each new birth with the flashing message: "More People Mean More Markets." The first popular credit card was issued in 1950 by Diners Club.

The pitfalls weren't shrouded in some mist of material bliss. Many critics offered lacerating judgments at the time. *Life* magazine reported on popular journalist William Whyte's study of spending by couples, in which he opined "they have hocked their incomes so far in advance that they are always strapped for cash." Journalist Vance Packard wrote an excoriating condemnation of planned, progressive, or dynamic obsolescence, in his 1959 number one *New York Times* bestselling book, *The Waste Makers.* Among a wealth of distressing revelations, he reported that marketing consultant Victor Lebow advocated "forced consumption" in an article for marketers, writing: "We need things consumed, burned up, worn out, replaced, discarded at an ever increasing rate."

Probably the most eloquent critic was one of the century's most eminent economists, John Kenneth Galbraith, who coined the term "conventional wisdom" and castigated the wisdom his profession had nurtured, writing in his bestseller *The Affluent Society,* "Few economists in recent years have escaped some uneasiness over the kinds of goods which their value system is insisting they must maximize. They have wondered about the urgency of numerous products of great frivolity." Single-use dog bowls certainly fit that bill. Training his ire on manufacturers and marketers, he asserted "the individual's wants . . . cannot be urgent if they must be contrived for him. . . . One cannot defend production as satisfying wants if that production creates the wants."

If recycling, repair, and reuse of products from companies that emphasized quality over quantity had enjoyed an equivalent boom, perhaps a planet-healthy balance of consumption and circularity could have been struck. But the view of recycling during the war as patriotic virtue, and the obvious economic choice, was quickly abandoned, and throwaways steadily eroded the demand for repair specialists. The duty to save and sacrifice had become a duty to spend and waste. The U.S. had become, as Lizabeth Cohen has dubbed it, a "consumers' republic."

The Grossness of Gross Domestic Product

Supporters of consumption as a means to prosperity lobbied economists and the government to glorify a previously little used economic concept called gross domestic product (GDP) as the core measure of economic health. The canonization of GDP sealed the deal for consumerism. Going forward, the performance of the economy would be measured by the increase in quantity of new goods manufactured, no matter whether they were of high quality, whether they harmed the planet, or whether they actually improved life in any appreciable way. The case for the duplicity and absurdity of GDP as the core measure of economic success was most clearly made by the economist who developed the metric, Simon Kuznets. In 1934, he wrote a plea to the U.S. Congress against using it as a gauge of prosperity because it disregarded so many aspects of the actual well-being of a society. "The welfare of a nation can . . . scarcely be inferred" by GDP alone, he cautioned.

Others tried to forewarn us as well. Galbraith, who would eventually be awarded the Presidential Medal of Freedom in 2000, wrote in *The Affluent Society* that "the thralldom of a myth—the myth that the production of goods . . . is the central problem of our lives," diverts public concern from social ills and environmental degradation. Arguing that true affluence is not a matter of higher consumption, he urged, "let us protect our affluence from those who, in the name of defending it, would leave the planet only with its ashes." The most eloquent critic of GDP may have been Robert Kennedy. In one of his most stirring speeches, given just days after he announced his candidacy for president, three months before he was assassinated, he lamented,

Our Gross National Product . . . counts air pollution and cigarette advertising, and ambulances to clear our highways of carnage. It counts special locks for our doors and the jails for the people who break them. It counts the destruction of the redwood and the loss of our natural wonder in chaotic sprawl. . . . It does not include the beauty of our poetry or the strength of our marriages, the intelligence of our public debate or the integrity of our public officials. It measures neither our wit nor our courage, neither our wisdom nor our learning, neither our compassion nor our devotion to our country, it measures everything, in short, except that which makes life worthwhile.

Why, then, did GDP become canonized as the official determinant of a nation's economic success? It occurred as Allied victory in World War II began to appear inevitable, during a single meeting of forty-four delegates of the Allied and friendly nations. They had convened for twenty-two days far from the front lines, at the elegant Mount Washington Hotel in the idyllic rural retreat of Bretton Woods, New Hampshire. Their aim was to agree on the creation of international economic standards and structures that would assure a global postwar flourishing after the devastation of the war.

The decisions made at Bretton Woods led to the building of some beneficial institutions, such as the World Bank and International Monetary Fund, which sought to provide financing for poor nations to develop. But it also indoctrinated economic measurement tools that drove economic superiority for the few and environmental degradation for all. With only men of similar ethnic, societal, and religious backgrounds attending, the decisions they made, whether intentional or otherwise, lacked a comprehensive understanding of how economies should function for the betterment of all. They agreed that the sheer total volume of goods an economy creates, whether built to last or to trash, energy hogging or energy conserving, life enhancing or life destroying, would be the single, stark metric of success. For example, while GDP may include rebuilding activity from a man-made disaster, it makes no admission of the cost caused by the disaster. It is analogous to rewarding a manager for fixing a problem he created without any note of the root cause of the problem, or any assurance that he won't repeat the error, and once again be rewarded for fixing it.

For example, GDP includes the manufacturing of toxic chemical fertilizers and drilling of greenhouse gas–belching oil and gas wells. However, it does not measure how clean our streets, waterways, and air are. It does not measure access to good health care. It does not measure access to quality education. It does not measure how safe our neighborhoods are. It does not measure the number of hours a person needs to work in order to earn a decent livelihood. Most oddly for an economic tool, it does not include the health costs caused by the chemical fertilizers and greenhouse gas emissions.

If we're to break free of the grip of consumerism, we will need to establish new standards that link the evaluation of an economy's health to the rise in the standard of living in a society. A number of viable options have been proposed. One is the World Happiness Report, issued each year, according to which the U.S., which ranks number one in GDP, ranked only number nineteen in the world. This goes a long way to explaining why the country is so riven by conflict and why so many Americans feel left far behind.

An alternative I'm particularly partial to is economist Kate Raworth's doughnut economics, which says that a country's economy should be assessed according to how well it is meets its people's life needs while also protecting its natural resources. Circularity is at the core of her model. In refuting the dictum that GDP must always increase, she minces no words about unbounded growth. "In our bodies, we call it cancer."

Raworth created this doughnut model to portray what national economies should aim for. The outer dark circle of the doughnut is the upper level of ecological damage that should be permitted. The inner dark circle is the threshold for providing the contributors to quality of life, below which a country should not descend. People must be provided at least that level of the robust set of life essentials she includes. In the middle is the doughnut sweet spot, or in Raworth's words, "the safe and just space" we should strive for. Vital to getting there, she argues, is building circularity throughout our economies.

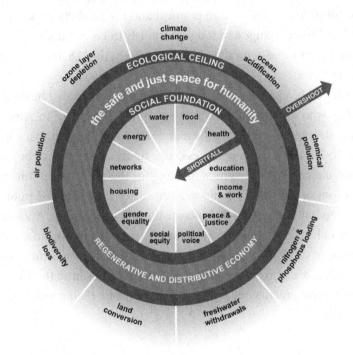

While the hope of abandoning GDP may well be quixotic, given that it's so entrenched, Raworth's doughnut economics is being taken quite seriously, not only by a number of leading economists, but by some government officials as well. The cities of Philadelphia and Portland have asked her to help them develop doughnut economic policies, and in the wake of the coronavirus outbreak, the city of Amsterdam announced that it was adopting Raworth's doughnut principles to set forth new goals for the restoration of its economy. "What we are looking at is how we can become a healthy and resilient city again," declared deputy mayor Marieke van Doorninck. "It gives us the opportunity to put other values—like social interaction, health, and solidarity—much more in the fore-front." How could any of us who've been through the pandemic not want the same?

But in order for politicians in the U.S. and much of the rest of the world to join in rejecting the gospel of consumption, the decep-tions of industry leaders and political and media abettors, who've denied climate change and fought furiously to keep extracting and polluting, must be neutralized. Unfortunately, the U.S. is a country of contradictions.

We've become a country in which many of the same business leaders who loudly caution about emerging "socialism" in Amer-ica, destroying the individualism and self-determination that sup-posedly led them to their success, also figured out a way to be the first recipients of government funding for their businesses during the onset of COVID-19 (only a decade after they required public funds to halt the imminent collapse of their companies). Authen-tic capitalism is a system designed to reward competition and merit; American capitalism has become a system that disproportionately rewards those who profess to be its adherents but behind the scenes have become practiced at gaming the economy for their own benefit.

2

The Disinformers

THE REQUESTS CAME FROM the Restaurant Action Alliance, reaching out to the owners of restaurants, delis, and food carts all around the city. Would they write a letter about the great duress their businesses faced? If they were too busy, no worries, they didn't actually have to write the letter—the Alliance handily provided sample text on its website. "This is just one more example of out-of-touch elected officials," owners should write, "who have no idea what it takes to run a small business in New York City."

Hecklers descended on a council hearing to denounce me. "Ron Gonen is a socialist," they yelled, "and he's turning Michael Bloomberg into a socialist too!" Our proposal was "just another Bloomberg Nanny initiative." Apparently moved by the opposition, City Councilman Robert Jackson declared at the hearing, "Quite frankly, I'm not sold on the administration's messaging. . . . I'm ready to recycle it instead of outright banning it."

A flurry of news articles warned of dire outcomes. An "industry report" had found that "a rough estimate" of the "impacts to the New York City region could be a net loss of around 2,000 jobs and $400 million in economic output." The effects would be felt well beyond the city too. Michael Durant, of the National Federation of Independent Business, called the proposal "a direct threat to thousands of jobs in upstate New York." So punishing would the proposal be on restaurants that owner Pablo Martinez, a reporter noted, predicted that "some owners may choose to close their restaurants and invest in another industry, or move to another state."

The cause of all of this clamor? As New York City's Deputy Commissioner for Sanitation, Recycling and Sustainability, I had submitted legislation to the City Council in 2013 to ban the sale of Styrofoam. It had become a financial and environmental menace. The city was spending close to $10 million annually to dispose of it in landfills, and when not captured in a trash can it caused severe environmental harm to our waterways, breaking down quickly into small particles eaten by fish and thereby also eaten by us humans too. It was also contaminating the city's recycling program, due to so many cups and so-called clamshell cartons being slopped with drink and food remains, which smear onto paper and cardboard and make it unrecyclable. Without the sellers of Styrofoam being willing to invest in an economically viable recycling solution, or agreeing to subsidize its disposal, Mayor Bloomberg and I, as stewards of taxpayer dollars and our local environment, decided a ban was the best option, especially since a number of recyclable and compostable alternatives were available, which many restaurants were already using.

For all these reasons, several other major cities had already enacted bans, including Portland (all the way back in 1990), Seattle, and San Francisco. Dire predictions had been made about those bans also, but the outcomes had defied them. Not reported in most of the stories about the predicted trauma for New York's restaurants was the fact that the ban included a waiver for any restaurant that could show it would cause them economic hardship. Requests for exclusion or financial assistance allowed for in the bans in those other cities had hardly ever been made. In Seattle, for example, out of forty-five hundred restaurants, only two applied for an exemption. Notably, years after the bans in Seattle, Portland, and San Francisco went into effect, their reputations as some of the world's great culinary cities continued to grow.

So why all the fuss? Because the largest producer of foam cups and clamshells, Dart Container Corporation, had a vested financial interest in New York continuing to purchase vast sums of their product. Dart has been a vigorous, and underhanded, critic of every Styrofoam ban proposed in the U.S. In California, as repeated efforts were made to pass a ban in the six years leading up to 2020, the company reportedly spent $3 million on ad campaigns and "donations" state legislators to try to defeat them. The National Resources Defense Council writes: "It is likely the movement to get rid of Styrofoam food and beverage containers would have proceeded more rapidly, were it not for an intense, well-funded industry disinformation campaign . . . led by the Dart Container Corporation."

The company's disinforming tactics can be quite elaborate. For example, the Restaurant Action Alliance, mentioned at the start of the chapter, described in press reports as "an organization composed of minority restaurant owners, managers and workers," and "a food industry lobbying group" was actually not an alliance of restaurant owners. It was an organization secretly formed and bankrolled by Dart, to the tune of $824,000. The actual New York Restaurant Association, the long-standing industry group that represents thousands of restaurants in the city and New York State, supported the ban.

Sowing confusion is the cornerstone of disinformation campaigns. In this case, the job was managed by Mercury Public Affairs, which describes itself as a "high-stakes public strategy firm." Not surprisingly, the "Alliance" didn't mention to the restaurant and food cart owners it riled up who its backer was. When asked to comment about the funding, Pablo Martinez, for example, who had predicted such hardship, said he'd had no idea, and added, "I feel a little confused." Another owner who said he was unaware of who was behind the Alliance added that he didn't use much foam but had simply "joined" the Alliance because he was "fed up with Mayor Bloomberg's policies surrounding public health and callous health inspectors."

It was eventually discovered that the wife of the Dart CEO made shadow contributions to a number of council members, including Robert Jackson, the city council member who fiercely opposed the ban. After leaving the council, he went on to register as an official lobbyist for Dart.

PERHAPS THE MOST BIZARRE ACCUSATION hurled during the campaign was that Bloomberg and I were thumbing our noses at a marvelous recycling opportunity. Headlines eulogized THE RE-CYCLING PLAN BLOOMBERG DOESN'T WANT, and decried that DE-SPITE RECYCLING PUSH FROM RESTAURANTS, MANUFACTURERS, NYC BANS STYROFOAM. A piece on the ban in the *National Review* asked, "Remember when nanny-staters tried to mandate the recycling of almost everything" and bemoaned "how the tide has turned," also oddly stating, "People who don't have a lot of money tend to use inexpensive convenience items, such as plastic bags, straws, and Styrofoam." So let me parse that. Trying to recycle everything was bad until it was good, and the middle class and wealthy don't use plastic straws, take home plastic bags, or ever order out? No mention was made in the flurry of complaints, of course, of who should cover the $10 million bill for landfilling foam the city taxpayers had been paying annually.

Eventually Dart proposed to the city that it would build a facility to recycle New York City's Styrofoam in the Midwest. Along with the curious concept of shipping the city's foam all that way to be recycled, buried in their proposal was that the arrangement would last only five years, after which the city would once again return to landfilling its foam. Despite the charade, they convinced a judge to overturn the ban in 2015. On appeal, a subsequent analysis led to the unanimous decision of a five-judge New York State Supreme Court panel, three years later, that the city had "rationally concluded" that Styrofoam "cannot be recycled in a manner that is environmentally effective and economically feasible."

After five years of underhanded machinations by Dart, and an additional $50 million in landfill and other costs to the city, the ban went into effect in 2019. Restaurant closings? The organization really representing the city's food purveyors, the New York Restaurant Association, had presciently stated in 2015, when announcing its support for the ban, that the economic effect on its members would be "nominal at best." Job losses in the city? Upstate? Not a peep about them in the news.

If the Dart family, which privately owns Dart Container Corporation, was so concerned about the public good, it might have instead lobbied the family scions Kenneth and Robert Dart to pay their taxes. They blatantly dodged millions of dollars of tax payments by renouncing their U.S. citizenship in 1994, to take advantage of a loophole in tax law. Kenneth, who was president of the company at the time, had the further audacity to request he be appointed a consul to Belize, where he had set up home, and be allowed to move into a consulate in Sarasota, Florida, where his wife and children had remained when he absconded. Working as a foreign diplomat would have gotten him out of a stipulation that as a noncitizen he could reside in the U.S. for no more than 120 days a year. President Clinton angrily denied the request, and at least the Darts's evasion inspired Congress to pass a law requiring those fleeing to pay an exit tax.

Insidious Tactics Galore

The antiban campaign was successful for so long because it employed a set of tactics right out of what the Union of Concerned Scientists calls the Disinformation Playbook. The entire repertoire of plays would require a set of encyclopedias to cover, and as Russia's disinformation machinations on Facebook show, they're constantly evolving. But we can dive into a particularly common, and potent, set.

Astroturfing: Create a bogus organization and give it a name that suggests it represents some beleaguered and mobilized citizenry, fighting mightily for their democratic rights—when in truth it's funded by and is fighting for interests in stark opposition. The term is owed to Texas senator Lloyd Bentsen, who in 1985 cleverly said about a manufactured write-in campaign sending a deluge of letters to his office in opposition to insurance regulation, "A fellow from Texas can tell the difference between grass roots and Astroturf."

Patients United Now, created to fight passage of the Affordable Care Act, claimed on its website: "We are people just like you." It was in fact run by Americans for Prosperity, founded and heavily funded by the billionaires David and Charles Koch. The group Washington Consumers for Sound Fuel Policy was created by the Western States Petroleum Association, funded by ExxonMobil, Chevron, BP, and other industry leaders. In response to Governor Jay Inslee's proposed bill to institute a carbon cap-and-trade program in Washington state, the group argued that the measure "could cost more than 11,000 jobs in the state"—a calculation from none other than the National Federation of Independent Business, the very same group that predicted that dire job losses in upstate New York would follow the foam ban. Inslee's bill was defeated.

Next tactic up: *shoot the messengers* (and for good measure, accuse them of being socialist "nanny-staters," or back in the day, communists). This tactic boasts a long legacy. A particularly sordid barrage was launched at Rachel Carson, author of the groundbreaking environmentalist clarion call, *Silent Spring*. Published in 1962, the book is a masterful exposé of the ravaging of our environment by the post–World War II boom in insecticide spraying, already well documented at that time by a host of scientific studies. So profligate was DDT spraying that author Charles Mann recalls, "Every spring tanker trucks rolled down our street, hosing down yards, trees and sidewalks with DDT. We kids followed along, shrieking with joy as the sweet-smelling, slightly sticky pesticide splashed over our faces and bodies." The book is known most for warning about possible large-scale bird death, the harbinger of coming silent springs, but Carson also spotlighted the devastation that had already occurred across a wide range of habitats and threats to many other species.

For that important public service, she was castigated in a $250,000 smear campaign (over $2 million in today's dollars) managed by the National Agricultural Chemists Association and paid for by the chemical industry. Monsanto distributed a brochure, "The Desolate Year," in mockery of the book's title, that portrayed widespread outbreaks of famine and epidemics of disease that would result if Carson had her way and pesticides were banned. Yet Carson hadn't called for chemical bans, neither of DDT nor any other toxin. Rather, she commended the advice of one specialist, who said "Spray as little as you possibly can" rather than "Spray to the limit of your capacity," and she concluded that longer term, "the ultimate answer is to use less toxic chemicals so that the public hazard from their misuse is greatly reduced." She went on to list a number of organic chemicals, derived from plants, that were already available. Hardly the "hysterically overemphatic" and "inaccurate outburst" the reviewer for *Time* magazine accused her of making, failing to note that she had been chosen over him for a job at the Fish and Wildlife Service years before. His review was titled, tellingly, "Pesticides: The Price for Progress."

The assassination was personal and appallingly sexist. Again and again Carson was demeaned as a "hysterical woman" and an "emotional female alarmist." A former secretary of agriculture, Ezra Taft Benson, wrote a letter to President Eisenhower in which he asserted that Carson was "probably a Communist," and wondered why a "spinster was so worried about genetics," referring to her coverage of the possibility that exposure to pesticides might lead to mutations. His implication seemed to be that a woman without children ought not have any concern for the children of others. The general counsel of chemical maker Velsicol sent a letter to her publisher threatening to sue, and suggested that she was working for the Soviet Union as part of a plot to create food shortages in the West. In truth, Carson championed democratization, asking potently about the spraying, "Who has decided—who has the *right* to decide—for the countless legions of people who were not consulted?"

Fortunately, so well crafted was *Silent Spring*, and so aware of nefarious, disinforming tactics was Carson, that she deftly overcame the onslaught. She wrote in the book that in response to "obvious evidence of damaging results of pesticide applications," the companies doled out "little tranquilizing pills of half truth." She was also already a beloved author by the time *Silent Spring* appeared, having written a trilogy of major bestsellers about the life of the seas, and she knew how to marshal her celebrity. In an hour-long appearance on CBS, she brilliantly put the polluters in their place, saying, "We still talk in terms of conquest. We still haven't become mature enough to think of ourselves as only a very tiny part of a vast and incredible universe." Velvet-gloved punch solidly landed.

So convincing was Carson, and so fierce was the public outcry over her revelations, that Congress quickly held hearings, with Carson as a witness. In 1963, just a year after the publication of *Silent Spring*, the Clean Air Act was passed. In the next decade, one after another of the toxins she had warned about were either banned or strictly regulated due to incontrovertible proof of their health and environmental effects. Unfortunately, Carson didn't live to learn of most of those actions; she passed away from breast cancer in April 1964.

Disinformers so often cling to clearly harmful business practices despite increasing regulation and consumer demand. If chemical companies had spent the past few decades pioneering green alternatives rather than pouring millions into duping the public and trying, to little avail, to hold on to their antiquated business models, they would surely have found an eager market among farmers, who hate that they've been poisoning their soil, and consumers, who obviously prefer products that don't harm their health. In their failure to serve public demand for clean products, they hurt the long-term interests of their shareholders. *They* were the anticapitalists.

The good news about the hypocrisy of nanny-state belittlements is that they can actually be easy to dispel, as Carson managed to do. One memorable experience I had while working for Mayor Bloomberg was with a guy who raced out of his house on Staten Island, the conservative bastion of New York City, to chew me out. When the agency I was running launched the curbside food-waste collection program, I joined the first team delivering the food-waste bins. Food waste represents over 40 percent of New York City's landfill expenditure—over $150 million annually—and eliminating that cost was one of our core initiatives. As I was leaning down to drop off a bin, the guy flung open his front door, bolted down the steps, and charged down the sidewalk at me. Jabbing a finger at my chest, he boomed, "I have *had it* with Bloomberg and his nanny state. He's using our tax dollars for his climate change gimmicks. They're bad for the economy and they're SCAMMING TAXPAYERS!"

The truth was that diverting food waste from landfills into either composting or anaerobic digestion would *save* taxpayers hundreds of millions of dollars. Mayor Bloomberg is a brilliant businessman, and he understood a fantastic economic opportunity when he was presented with the curbside food-waste collection program. For the prior decade, NYC had been spending over $300 million of taxpayer money to transport its waste to landfills in Pennsylvania, Ohio, and South Carolina. Bloomberg understood that all of that so-called waste should be seen as a treasure trove of valuable commodities. Masses of aluminum, cardboard, and plastic could be sold for good money. Food waste, the largest component of the city's trash that was being transported to landfills hundreds of miles away, could instead be converted to rich compost and sold to landscapers or turned into natural gas via the entirely natural process of anaerobic digestion. That gas could be used to run the city's massive fleet of garbage trucks, saving still more costs. The city would earn a handsome profit and the increased recycling activity would create jobs.

Such is the beauty of circular economy solutions; they are superoptimal, providing critical solutions while producing financial rewards for both businesses and the public. In short, they use capitalist opportunities to protect our environment. Which is why when I explained to my Staten Island adversary that the city was spending over $150 million every year to send *just its food waste* to landfills, and that the food-waste collection program could eliminate that cost while also creating thousands of local jobs in the composting and anaerobic digestor industries, he looked surprised. Then he took a moment to absorb and finally said, "Wow, I wish someone would have explained that to me years ago. Give me that bin."

So often, disinformers become more intent on digging in, rather than seizing planet-healing business opportunities, even as evidence of the damage steadily mounts. Rather than heed the evidence, they opt to discredit it. Hence the popularity of another of the playbook's tactics, *wolf cries wolf*: accuse your opponents of exactly the deception you are perpetrating. "There's so many lies being told," a spokesperson for Dart told the press, "so we had to engage and be part of the political process." Alan Shaw of Plastics Recycling, Inc., the company Dart cooked up its five-years-only deal with, said, "Once again, New York City is ignoring the facts that prove polystyrene foam can be recycled," We see the tactic from climate change deniers defending the fossil fuel industry when they complain that people only have access to information that fossil fuels cause climate change because science journal editors only publish submissions that support their belief in climate change. But as Jane Mayer exposed in *Dark Money*, reams of books and journals are sponsored and published by climate change deniers.

The professed indignance can reach comical extremes. Take the fevered complaint by a spokesman for the Advertising Mail

Marketing Association that "for decades misinformation, disinformation, and outright lies regarding our industry have been broadcast. . . . The term 'junk mail' has been used and popularized by newspapers as a weapon to disparage our industry," which "presumes that advertising mail is without value, merit or quality." If junk mail had any merit, then American households would not be discarding an estimated 5.6 million tons of junk mail every year in the U.S., with most of it ending up in landfills. It makes sense that environmental and consumer affairs groups have long sought to ban it.

Because such professions of grievance by perpetrators often appear self-serving, a more clever approach is to have supposedly unbiased compatriots make your case for you. This is the tactic the Union of Concerned Scientists dubs *the screen*: funding institutions and, through grants and fellowships, the work of individual scholars who produce dubious, often outright tainted, research that supports their cause. Organizations founded for this purpose are typically given innocuous names that mask their fealty, such as the Foundation for Research on Economics and the Environment, which argues against environmental regulations, and the Pacific Research Institute, which has promoted climate change skepticism and worked against the plastic bag and Styrofoam bans in California. To imply that such groups are working in the broad national interest, a nifty sleight of hand is to simply dub them "National" something, such as the National Center for Public Policy Research, as though they are federal government organizations. All these groups have received substantial funding from ExxonMobil and other fossil fuel giants, as well as from think tanks opposed to environmental regulation, which in turn receive ample funding from fossil fuel and mining companies.

The funding of work by sympathetic scholars within mainstream academic institutions is a particularly shameful variation of this tactic, one that Rachel Carson warned about over fifty years ago. She wrote about "certain outstanding entomologists" who supported massive spraying. "Inquiry into the background of some of these men reveals that their entire research program is supported by the chemical industry. . . . Can we then expect them to bite the hand that literally feeds them?" So too for the many academics who have contributed to climate change denialism, often with financial support from the fossil fuel, mining, and chemical industries.

A primary player in promoting bogus denier science has been Exxon, as glaringly revealed in a 2019 report titled "America Misled." Written by a group of Harvard scientists, the report was submitted as evidence in a lawsuit brought that year against Exxon for committing fraud against its shareholders. According to internal company memos published in the report, Exxon was aware by 1977 that, quoting from its memo, "CO_2 release" was the "most likely source of inadvertent climate modification"; they understood the potential warming effects of emissions as early as the 1950s. The memo predicts that global temperatures could increase by 1 to 3 degrees by 2050, and by a staggering 10 degrees at the poles.

Yet despite this awareness, the company opted to dig in on oil and gas extraction—really, really deep—accompanied this with their launch of a decades-long disinformation campaign. A 1988 internal memo lays out the strategy in bold black-and-white. The company would "emphasize the uncertainty in scientific conclusions," though as the authors of "America Misled" point out, by that time the scientific community had reached a strong consensus that climate change was happening and that it was being driven by human activity.

The Union of Concerned Scientists calls this tactic *the diversion*, citing a memo written by a tobacco company executive who spelled out in stark terms how it works: "Doubt is our product, since it is the best means of competing with the 'body of fact' that exists in the minds of the general public." As for Exxon, it declared in a 1989 memo that it would have achieved victory in its disinformation campaign when "average citizens 'understand' (recognize) uncertainties in climate science" and "recognition of uncertainties becomes part of 'conventional wisdom.'" Mission achieved. So effective have been the company's efforts, and those of the horde of other deniers, that it's taken the devastating weather extremes and the intensifying melting of glaciers and polar caps to convince the majority of the public that emissions-driven climate change is all too real, and all too rapid.

Why Not Innovate Instead?

One of the foundational arguments for the superiority of capitalism is that it incentivizes innovation. Yet the extractive industries that have portrayed themselves as great defenders of free-market capitalism have stifled innovators in order to eliminate any competition to their antiquated and pollutive business models. They've worked furiously to artificially prop up their profits, at taxpayers' great expense, via lobbying for, and receiving, enormous tax subsidies. They've caused enormous damage to publicly owned environments, rather than develop green alternatives, an industry that is now booming. They have also worked hard, with their lobbying and public propaganda, to thwart the innovation of superior green solutions by others.

In 2019, wind and solar energy were the fastest growing sector of the energy industry. Stock prices of renewable energy and electric vehicle companies continue to rise, while Exxon's stock continues to decline. The anticompetive and disinformation tactics that Exxon executives and its board employed for decades have destroyed tens of billions of dollars of shareholder value.

Again, Rachel Carson warned us about the gross inadequacies of this model back in the 1950s. When the use of pesticides in agriculture spiked, while farmers in the developing world were facing a low-yield crisis due to pests and drought, in U.S. the situation was quite the opposite. "We are told," she wrote, "that the enormous and expanding use of pesticides is necessary to maintain farm production. Yet is our real problem not one of *overproduction*?" At the time, taxpayers were funding more than $1 billion in annual subsidies to purchase excess crops and pay farmers *not* to plant. Farmers have received subsidies ever since, reportedly rising as high as $32.1 billion in total in the year 2000, and coming in at $22 billion in 2019.

A particularly egregious case of opting not to innovate involves none other than Exxon.

The company foresaw by the early 1960s the possibility that its oil business would become untenable because of fierce competition from the Middle East. Executives began contemplating, in the words of an engineer working for the company at the time, "What if these producers start jacking up the price and our market dries up?" and "What can we do if we can't be in the oil business at all?" Out of that concern, the company became the earliest commercial backer of solar power development, a smart, innovative move. It even installed solar panels to power its oil drilling platforms in the Gulf of Mexico. By 1973, the Solar Power Corporation, funded by Exxon, was selling solar panels all around the world. Yet, despite the fact that repeated oil shocks were inflicted on the global economy by Middle Eastern producers in the 1970s, underscoring the value of alternative energies, by the middle of the 1980s Exxon abandoned its solar business because it had determined it would take until 1994 or so before it would be able to stand on its own profitably. How remarkably shortsighted. Given the current solar boom, it's clear that the behemoth walked away from an enormous economic opportunity.

We can take comfort from the fact that these duplicitous strategies generally lose over time. Unfortunately, as we are learning with climate change, the human and economic toll of that war can take generations to fully recover from. Renewable energy, despite industry pushback, has reached a tipping point, with the cost of solar production having dropped so precipitously in the past decade that solar power is now cheaper in many markets than oil. Meanwhile, the costs of oil and gas extraction have inexorably increased, and oil company profits have been severely pinched. The harsh, capitalist truth of the oil and gas industry's business is that it's a poor investment, hyperdependent on government subsidies and the public allowing it to spoil its land without consequence.

As Naomi Klein highlighted in her book *This Changes Everything*, companies are going to extremes to tap new oil reserves, such as with ultra-deepwater "subsalt" extraction, which pulls oil up from depths of up to ten thousand feet. At the same time, the boom in the methane-leaking process of extracting natural gas through hydraulic fracking has led to such oversupply that according to an energy industry assessment, the business "could be headed off a financial cliff." Smaller oil and gas producers have been so challenged that in the past four years an estimated 175 of them have filed for bankruptcy protection. As for the giants, Exxon's profits were cut in half in 2019 and its share price has stayed virtually flat for the past decade.

Note that the companies are in these financial straits despite the fact the industry has been subsidized by taxpayers for more than a century, starting in the 1910s. Estimates of the total annual subsidies vary, but one source puts them at $10.7 billion in 2019 for the U.S. alone. The fossil fuel lobby has cried foul about "socialistic" subsidies to the solar and wind industries, which are estimated at about a fifth of those the fossil fuel industry itself welcomes; and solar and wind subsidies in the U.S. are scheduled to end in 2020. The mission has been accomplished; the green industries are predicted to continue growing at a good clip due purely to free-market demand. Subsidizing nascent businesses that are seen as in the national interest makes good economic sense, but subsidizing large, antiquated businesses for decade after decade does not.

In *Dark Money*, Jane Mayer quotes William Simon, then the president of the Olin Foundation, one of the organizations funded heavily by the extractionists to promote denialism, complaining that "capitalism has no duty to subsidize its enemies." He was referring to subsidies to universities, which he saw as hotbeds of hostility to capitalism. I would counter that capitalism has no duty to subsidize struggling businesses that have failed to innovate into the greener terrain that the public desires. Unfortunately, the oil companies did move into new lucrative terrain, but it was anything but green; they pivoted into virgin plastic production.

Plastic Denialism

Most people probably assume that the oil and gas industries make their money from selling fuel. But 99 percent of plastics are made from petroleum and natural gas chemicals, and turning fuel into plastic stock has become an increasingly large portion of the oil industry's business. This explains why Exxon, Chevron, Shell, and company have contributed to the fight against the host of proposed plastic bag bans making their way to state legislatures in recent years. In fact, it was Exxon that brought the plastic grocery bag to the U.S. in 1976. In market tests, neither grocers nor consumers were at all pleased with them, despite their being printed in red, white, and blue in honor of the bicentenary. Many shoppers outright hated them, as one grocery clerk attested, telling a reporter that "some customers become real irate and start shouting." Convincing stores to switch from paper to plastic took quite a PR effort over more than a decade, with some laughable salvos. One press release, for example, suggested plastic bags were wonderful because they could be repurposed in seventeen different ways, including "as a jogger's wind breaker or a beach bag." So intent was Exxon to make them catch on that the company devised a plethora of designs for bag holders to make them easier for clerks to load. Their adoption was, in short, hardly a matter of free-market demand.

The campaign against bag bans has studiously followed the Disinformation Playbook. Start an organization with a name that implies it's serving the public good: The American Progressive Bag Alliance. This so-called organization, spearheaded by the American Chemistry Council, listed its address on the Alliance website as "PBA c/o Edelman, 1500 Broadway, New York, NY 10036." That's Edelman, a PR firm. A lynchpin of its persuasion efforts has been a false dichotomy: plastic bags cause less environmental damage than paper bags. The reality is that paper is easily and profitably recycled while municipal recycling facilities usually struggle to find markets for plastic bags. Worse, plastic bags often get wrapped around the expensive sorting machinery. This means that recycling facilities are forced to shut down for hours at a time to clean out a 2-cent plastic bag that had become entangled in a $500,000 piece of machinery. Most egregious, though, is the false equivalence being perpetuated. For most of history, people shopped with reusable bags or carts. It's hard to imagine a more inefficient and costly system than one that expects either the retailer or the consumer to pay the cost of new bags every time they shop.

The Media Abettors

The disingenuous tactics of the disinformers are often not hard to uncover. Any reporter wishing to do so can usually find good data about any given organization professing to serve the public good, including where its funding comes from. While it's shameful that corporations and their PR and think-tank collaborators launch disinformation campaigns, it's clear why they do it. Harder to fathom is why the mainstream press sometimes perpetuates their falsehoods. As I became acutely aware while in charge of recycling for New York City, the media has done so with particular zeal when it comes to recycling. For decades, the media has forecast its demise, even as the industry has grown, in the U.S. alone, into a $117 billion powerhouse and provider of 534,500 jobs.

Throughout 2018 and 2019, a barrage of headlines in major newspapers and network news coverage screamed "Who Killed Recycling?" and "Why America's Recycling Industry Is in the Dumps." The cause of the doomsaying? In March 2018, China implemented its National Sword policy, declaring that it would no longer accept masses of the many recyclables it was importing, in particular contaminated paper and low-quality plastics. The volume of material that had been sold to China was staggering. By one report, the second largest waste collector in the U.S., Republic Services, sold 35 percent of its total recyclables to China in 2017. That number dropped to 1 percent in 2018.

Reports of the dire straits many recycling programs were facing were certainly sobering. *The New York Times* reported that "hundreds of towns and cities across the country . . . have canceled recycling programs, limited the types of material they accepted or agreed to huge price increases," referring to fees charged by the recyclables collectors. CBS News announced that "mountains of paper have piled up at sorting centers, worthless. Cities and towns that once made money on recyclables are instead paying high fees to processing plants to take them. Some financially strapped recycling processors have shut down entirely, leaving municipalities with no choice but to dump or incinerate their recyclables." Yet, when a spokesperson at the country's largest recycling company, Waste Management, which controls about half of the U.S. market, was asked in March 2019 by a reporter for *Resource Recycling*, a leading trade journal, about how many cities had halted their programs, the answer was, "Of our over 5,000 municipal contract customers, we have only identified two that have chosen to pause or stop their recycling programs to date." Even more to the point, not a single one of the top twenty-five largest U.S. cities eliminated its recycling program.

Most programs that were stopped were those in smaller and rural communities. The correct characterization of the situation is that it was a disruption, which ultimately led to important and long overdue innovations in technology and the development of domestic markets for recycled materials. Furthermore, it was a recognition by the Chinese of the value of the recycling industry and a desire on their part to become a major economic player in the industry. Beginning in 2018 the news site/online publication Waste Dive made a comprehensive effort to track how many programs in all fifty states were discontinued. While the effects have been widespread, they found the number much smaller than many had claimed. Out of over ten thousand curbside recycling programs, a little more than 100 were canceled by the start of 2021, none in major cities.

Most news coverage didn't highlight the marvelous advances in sorting technology made in the last decade, such as optical sorters that use infrared light to identify different types of plastic and air jets specifically programmed to blow those different types off a conveyor belt and into specific bins. Recently, robotics and artificial intelligence have been introduced into recycling facilities, increasing yield, margins, and reporting.

The host of doomsaying articles also didn't reveal that in response to China's ban, $1 billion was invested by private companies and investors in U.S. paper mills to expand recycling in just the first six months of 2019. China has been a leading investor. One company alone, the Hong Kong–based Nine Dragons, the largest recycler and manufacturer of cardboard boxes in the world (owned by the wealthiest woman in China), poured a reported $500 million into either reviving shuttered paper mills or expanding running ones. In addition to creating paper products for sale in the U.S., Chinese firms are turning paper waste they formerly imported into pulp here instead, and then importing that. U.S. and Australian companies are investing heavily too, including Pratt Industries, which is building a big new plant in Ohio for processing recycled paper and turning it into boxes.

Substantial new investment in plastic recycling facilities is also under way. In an article published by the Sierra Club in June 2019, the president of the Association of Plastic Recyclers said, "The whole crisis narrative has been wrong. China didn't break recycling. It has given us the opportunity to begin investing in the infrastructure we need in order to do it better." The correct takeaway about the ban is, in short, that it has showcased the enormous business potential of developing domestic markets, which should have been going on in the first place instead of exporting to China.

It's important to recognize that some in the news media have done a great deal of balanced reporting on recycling through the years, but there are some reporters who have recognized that a negative story about recycling, regardless of its merits, drives eyeballs—which is, too often these days, the measure of a success-ful article, unfortunately. Probably none has done so more effectively and notoriously than John Tierney, who in 1996 as a reporter for *The New York Times* wrote a cover story for the *New York Times Magazine* titled "Recycling Is Garbage." The article garnered major coverage and mainstreamed a number of myths and distortions championed by ardent think-tank recycling foes, quoting scholars from the Cato Institute, the Reason Foundation, and the Competitive Enterprise Institute, all of which receive funding from petrochemical companies and are identified by Jane Mayer as leaders of the dark money nexus.

Tierney's piece reads as almost tongue in cheek. When he ridicules an elementary school project that opened kids' eyes to the value of resources in trash, he argues, "Mandatory recycling programs aren't good for posterity. They offer mainly short-term benefits to a few groups—politicians, public relations consultants, environmental organizations, waste-handling corporations—while diverting money from genuine social and environmental problems." He makes no mention that landfilling—the alternative he enthuses about—benefits the waste haulers and landfillers at substantial public expense. Indeed, nowhere in the piece does he address the considerable fees the public pays for landfilling, figures which are readily available.

He defends plastic packaging because it takes up less room in landfills than other materials, even though, as he strongly emphasizes, supposedly there is no problem at all with finding enough space for landfilling (about which more in a bit). He also makes no mention of plentiful studies by that time revealing that plastics and their toxins were polluting rivers, lakes, and oceans. Regarding the value of conserving forests through recycling, he asserts that "acting to conserve trees by recycling is like acting to conserve cornstalks by cutting back on corn consumption." But of course, we don't harvest the leaves of trees and leave their trunks; it's the trunks we use, and forests perform immensely important carbon sequestration.

One of his most absurd assertions in the article, which has been hauled out again and again in the media, is: "If Americans keep generating garbage at current rates for 1,000 years, and if all their garbage is put in a landfill of 100 yards deep, by the year 3000 this national garbage heap will fill a square piece of land 35 miles on each side." This has become a meme, parroted again and again by antirecyclers, usually with some fuzzifying math. In one case the hole, it was said, would be "44 miles wide on each side and 120 feet deep," in another "15 square miles in size," with no depth cited. Not to be outdone, contrarian provocateur John Stossel states in a video titled *Green Tyranny*, "You could put *all the world's trash* for the next thousand years into one fifteen-square-mile landfill."

What is the original source of the meme? Professor Clark Wiseman published it in a paper in 1990, in which he was identified as being affiliated with Gonzaga University. Not specified was that he was also a fellow at the Property and Environment Research Center (PERC), which has received funding from Exxon, the Koch Foundation, and other extractive concerns, and has steadfastly promoted antirecycling myths as well as climate change denialism, even publishing a 2007 report titled "The Benefits of Climate Change." For the record, the space for landfilling has in truth been limited in many states, particularly in the Northeast and in California, and the costs of landfilling, which Tierney sidestepped, are considerable. In the simplest terms, do you know anyone who wants to live near a landfill or pass one on the way to and from work every day?

Probably the most egregious claim Tierney made is that "recycling has become a goal in itself," of no other value, completely ignoring that the goal is to help heal the planet. Journalist Chris Mooney writes that a reported friend of Tierney's, conservative writer Christopher Buckley, described him as "a bit of a merry prankster," but publishing such a disingenuous piece in the nation's leading newspaper was no mere prank. When Mooney asked Tierney about the piece and if he were "an equal-opportunity debunker,"

Tierney responded, "I could write something about the good side of recycling. . . . But everybody else writes that." Hardly. One wonders whether Tierney's editors at the *Times* asked him the same question, and why, given the outpouring of criticism of the piece, including a point-by-point rebuttal by the Environmental Defense Fund, the editors allowed him to write a reprise in 2015, in which he asserts that since the first piece was published, "While it's true the recycling message has reached more people than ever, when it comes to the bottom line, both economically and environmentally, not much has changed at all." In fact, the economic value for municipalities and investors created by the recycling industry has grown exponentially. What hasn't changed at all is the massive economic and environmental cost of sending waste to landfills.

The recycling industry has had a similar experience to most commodities-based businesses during the past few decades. In 1992, after years of increases in the value of recycled paper attracting new entrants into the market, the industry saw a massive increase in the supply of newspapers for recycling, which drove the price of recycled paper down. The market eventually corrected, and paper recycling became highly profitable again. By the early 2000s, recycled paper had become America's number-one export, with most of it going to China, which the country used to support their manufacturing boom. In 2009, another shock hit, due to the general economic downturn caused by the 2008 financial crash. As we'll see in the chapters ahead, recycling businesses that are truly devoted to recycling as their mainstay business, as well as China's National Sword policy, have generally weathered these challenges and have prospered. Much of the confusion over recycling comes from it being written about as one business, when in fact there are many recycling businesses—for paper, for plastic, for metals, and for glass. Some are considerably more challenging than others, but all are viable, and in each, as we'll see, breakthrough innovations are rapidly emerging. Most important to municipal leaders is to recognize that the value of a recycled commodity is secondary to the savings generated by avoiding sending that commodity to a landfill. For example, assume the value of a recyclable commodity was $0 per ton. Not recycling it, and therefore sending it to a landfill in the United States, would cost the municipality on average $50 per ton. While recyclable commodities generally have value—and in the case of commodities like aluminum, significant value and margin—one should always start their economic analysis from the standpoint of "If I don't recycle it, how much will I have to pay to landfill it?"

The good news is that innovations in reducing the amount of resources extracted, products designed for longevity and repair, and models for product and packaging reuse are starting to flourish. The forces advocating circularity systems and business models have begun to win the information war as well, because they are proving, as Paul Hawken foretold more than twenty-five years ago in his seminal circularity manifesto *The Ecology of Commerce*, that we can create a new system of production and consumption "that is so intelligently designed and constructed that it mimics nature at every step, a symbiosis of company and customer and ecology."

The final question that no conversation or journalistic investigation about recycling should ever ignore is, Who pays? Not who pays for recycling, but who pays if we don't recycle? If we stopped recycling in the U.S., American cities would need to come up with over $5 billion annually to landfill refuse, would lose hundreds of millions of dollars in revenue from the sale of material, and would lose tens of thousands of local jobs in the recycling and manufacturing sectors. They would also have to determine whose neighborhoods would bear the loss to their property value of having additional landfills sited close by. Things that can't be recycled, and things that can be recycled and are not, have a direct and extreme economic and environmental cost—like the effects of cigarettes on your health. And like Big Tobacco, the disinformers—whether it be the makers of DDT, the Dart Container Corporation, or ExxonMobil—won't give up without a fight, one that consumers are set up to lose.

Circularity Innovators Forge Ahead

W HEN I MET ALGRAMO FOUNDER José Manuel Moller, I knew that our missions to advance circular business models were aligned. The similarities in our backgrounds were actually a bit uncanny. Like me, José had decided to become an entrepreneur while in business school in Chile, where he grew up, and he was just as skeptical of the economic dictums that he was being taught about the actual existence of free markets. He told me that the "Chicago Boys"—a reference to the hard-line, free-market economic theorists at the University of Chicago who led the charge for deregulation in the 1980s, and took laissez-faire arguments to new extremes—had "experimented" on the Chilean economy. As he put it, "Chile was their laboratory."

Acting as advisers to the brutal dictator General Augusto Pinochet, Milton Friedman, Arnold Harberger, and colleagues helped the Chilean government orchestrate an economic "shock treatment," as they described it, that was supposed to unleash the marvelous energies of deregulation and power growth—but that in fact caused great economic pain. Inflation soared, with prices for consumer goods rising by 375 percent on average the first year, and employment plummeted.

José saw the devastating effects on working-class Chileans when he moved into the Recoleta neighborhood of Santiago while in business school. As I had seen with my mother's struggle, he saw that the people in the community worked hard all day but were barely scraping by. Shopping in the small bodegas that are the economic heart of the neighborhood, he was deeply moved by the stamina and business savvy of the women who run 95 percent of the stores, while also raising families. As the one in charge of buying all the household items for himself and his two room-mates, he realized that the people in the community were paying a hefty "poverty tax." Because they couldn't afford to purchase in bulk, they were paying as much as 50 percent more for all sorts of goods, from rice and beans to laundry detergent. And so his inge-nious idea for Algramo was born. He decided he would start a com-pany that would enable customers to purchase goods in smaller quantities—by the gram (hence the name Algramo)—at much lower prices, due to the elimination of the cost of packaging.

Reflecting José's commitment to the environment, this solution would also tackle the problem of the plastic packaging waste heaped in small garbage dumps all around the neighborhood. If he sold products in refillable containers, after the initial purchase of a container, the customers would buy only the amount of product they wanted. With packaging accounting for as much as 40 to 50 percent of the cost of many of the items in smaller packages, that alone would allow him to bring prices way down. In addition, he devised the idea of "packaging as a wallet." By implanting Al-gramo's containers with an RFID chip, a program could keep track of each time customers refilled, it enabled the company to offer a 10 percent discount for each refill. That way, with each purchase, they would earn back the cost of the container. He found a local manu-facturer to make the containers—another way he could contribute to the local economy—and he made purchasing easy by offering refills through a fleet of mobile dispensing units mounted on electric tricycles. He also partnered with two thousand bodegas throughout Santiago to install dispensing machines in their stores.

So innovative is his model that he won a coveted MIT Solve award for entrepreneurs with potentially transformative solutions for social problems, which helps scale their enterprises up. MIT scientists are now helping José develop his data analytics capabilities so that over time he can continually refine his offerings and also assess Algramo's positive environmental effects. In 2020, *Fast Company* named Algramo the "Most Innovative Company in Latin America." Recently, in partnership with Closed Loop Partners, Algramo expanded beyond Latin America and launched service in the U.S. with installations of dispensers in Brooklyn.

Ideas That Can't Help but Change Your Mind

José told me he learned about circular economics through friends in college who were environmental activists. For my part, I was fortunate to be initiated into the movement by my friend and mentor, architect Paul Macht. He was an early practitioner of green building, crafting passive solar houses for clients well before they gained any popularity. Paul has a great depth of knowledge about the roots of the movement and the key thought leaders and implementers who've made seminal contributions. Paul was also my high school water polo coach, having been a star water polo athlete himself. In my sophomore year, when my single mother was ill and confined to the hospital, I was incredibly fortunate to spend a good deal of time with Paul after he invited me to move in with his family. On forty-minute drives to and from practice, he shared his passion for innovation in environmental protection and restoration. I also witnessed his dedication to solutions, as he transformed the old farmhouse where he had just moved his family into a state-of-the-art green home. He even repurposed a large chicken coop on the property into his office. Why tear down a perfectly sound structure?

I kept in touch with Paul after I headed off to college, and in 2002 he gave me a copy of the newly published, groundbreaking book *Cradle to Cradle* by architect William McDonough and chemist Michael Braungart. For me, as for so many, it was enormously inspiring. McDonough and Braungart laid out a powerful case for designing products so that the materials they're made from can be easily repurposed. Though McDonough and Braungart popularized the term "cradle to cradle," they didn't coin it. That was done by another architect, Walter Stahel, the Product-Life Institute founder.

Stahel came up with the term in the late 1970s, out of annoyance with a great deal of buzz at the time about "cradle to grave" disposal of hazardous waste. Legislation had just been passed in the U.S. that made companies that produced toxic waste responsible for safely disposing of it in "graves" that would prevent any leakage. What a pathetically limited vision of what human ingenuity is capable of, Stahel thought. Wouldn't it be better to not produce hazardous waste at all, or any waste; why not manufacture in ways that allow for products to be "reborn" in circular loops? Stahel and McDonough are among a number of other architects who were joined by visionary economists and ecologists in helping make the business case for circularity. As the disastrous effects inflicted on Chileans have demonstrated, economics has done much to earn its moniker as "the dismal science." But some renegades in the profession made major contributions in the 1960s, '70s, and '80s to the understanding that an economy can, and should, prioritize circular systems and business models.

They raised the alarm decades ago about environmental degradation, inspired by the shocking realization that the planet's resources were already so depleted that before long, without dramatic transformation of the take-make-waste economy, Earth might no longer be able to sustain life.

Portrait of a Pale Blue Dot

One of the greatest achievements of human ingenuity of all time was the mission of Apollo 8, which opened the whole world's eyes to the planet's fragility.

In 1968, standing thirty stories tall, the Apollo 8 Saturn V rocket was by far the most powerful spacecraft constructed to that date, sent on the most audacious mission yet attempted—to break free of Earth's firm gravitational grip and propel itself into orbit around the Moon. The craft would need to travel 240,000 miles and reach an unprecedented speed of precisely 24,200 miles per hour to burst through the Kármán line, the barrier between Earth's atmosphere and space. It would then need to modulate its speed to 5,000 miles per hour in order to approach the Moon at exactly the right moment to fling itself, as it approached the Moon's dark side, into orbit with one last engine burn in a harrowing maneuver NASA dubbed "translunar injection." If the craft's single engine failed during injection or the astronauts miscalculated the burn by even a second too soon or too late, they and their tiny module would spiral out of control, either dashed to bits on the Moon's surface or catapulted out of orbit into a free fall into outer space from which there would be no return.

Initially slated only for Earth orbit, the mission had been abruptly upgraded in the hope that the U.S. could outrace the Soviets in flying to the Moon. The three astronauts, Frank Borman, James Lovell, and Bill Anders, were put through physically intense and mentally grueling training paces. So dangerous was the mission that when Anders's wife asked mission chief Chris Kraft what he thought the odds of the astronauts' safe return were, and he answered 50 percent, she was relieved—but a great deal more so when, in a remarkable engineering triumph, the lunar module reached the targeted spot for catapulting into orbit at precisely the calculated second.

A prime objective of the mission was to photograph the lunar surface, scouting out viable sites for a later Moon landing. On the module's fourth rotation, as Borman gently turned the module to a new angle, the astronauts saw the lunar horizon loom before them, the Moon's curvature of gray set against the deep black vastness of space. And suddenly, peeking up over the horizon, rose a tiny bright orb of blue and white.

"Oh my god!" Bill Anders shouted. "Look at that picture over there! Here's Earth comin' up! Wow, is that pretty!"

For the first time, humans had seen our earthly oasis in its entirety, its lapis blue oceans and billowy white swirls of cloud cover hovering in the limitless expanse of space as if a beacon to guide the three intrepid travelers safely home. Frank Borman recalled thinking, "This must be what God sees." Anders hurriedly captured a photo known as Earthrise, which is said to have had a transformative effect on humanity's consciousness of how finite Earth's precious resources are. As Robert Kurson wrote in his account of the mission, *Rocket Men*, "The astronauts had come all this way to discover the Moon, but they had discovered the Earth."

Two years prior, English economist Kenneth Boulding had published a highly influential essay titled "The Economics of the Coming Spaceship Earth." As environmental scientists had developed a broader and increasingly refined understanding of the magnitude of resource degradation, species endangerment, and atmospheric damage, the concept of Earth as comprised of delicately balanced ecosystems had come into focus. That consciousness drove a call for a radical new economics—an economics modeled on nature's circularity, with no resources wasted, in an ecological cycle of renewal.

Boulding contrasted what he called this new "spaceman economy" with the industrial "cowboy economy," which had been "reckless, exploitative, romantic, and violent." In the new economics as Boulding envisioned it, "all outputs from consumption would constantly be recycled to become inputs for production," and Earth would be understood "as a single spaceship, without unlimited reservoirs of anything, either for extraction or for pollution, and in which, therefore, man must find his place in a cyclical ecological system."

With the Earthrise photo, we had seen the spaceship. The environmental movement was galvanized, and a number of pioneering thinkers began the work of developing the circular-loop economy Boulding imagined.

Economy as Ecosystem

One of those visionaries is Herman Daly. He popularized the notion of what he called a "steady-state economy," which, he wrote, is "by no means static" but is one of "continuous renewal." Reading *Silent Spring* was his inspiration. He recalls realizing, in a shock of insight, that "once you sit down and draw a little picture of the economy as a subset of the larger ecosystem," *all* economic growth was coming at the expense of the ecosystem. We had foolishly come to think of the human economy as independent of nature, he argued, as a man-made mechanism, when in fact it is utterly reliant on the natural world and has a great deal to learn from nature's processes. He went on to promote a "radical shift from a growth economy"—and the idolatry of GDP—to a sustainable economy, and became one of the primary champions of sustainable development.

The vexing question was, how could a new economy that preserved, or even better, replenished the natural world, still be a thriving economy, one that allowed us to keep improving the quality of life—including for the developing world? In seeking answers, the economists turned to ecology, and Daly was highly influential as one of the founders of the new field of ecological economics, which laid the groundwork for the concept of circularity.

While various accounts of the evolution of the idea of the circular economy differ in detail, there is no dispute that its fundamental premise was that the human economy should not only respect nature but also reflect nature. We could make so much better use of natural resources by following the lead of nature's own systems of production, as well as its systems of destruction and reproduction. Ecologist Barry Commoner pointed the way specifically to the notion of circularity in his own bestselling 1971 book, *The Closing Circle*. He persuasively described the essential superiority of the natural economy, coining a maxim that became a core driver of the development of circular economics: In nature there is no waste.

That doesn't mean there isn't plenty of trash—an insight pointed out to me by biologist and biomimicry expert Dayna Baumeister. When you think about it, all sorts of natural trash come to mind. Many critters are profligate litterers. Snakes slither out of their skins and leave them strewn about forest floors. Squirrels manically spit out flakes of acorn skin as they chew down to the luscious, fatty core. As for trees, they toss their leaves as soon as they have no further use for them, though Baumeister explained that, in a brilliant demonstration of nature's ecological wizardry, they first suck out the nitrogen, a vital tree nutrient. That, in fact, is the reason leaves turn brittle and brown.

The thing that distinguishes natural waste from human is, of course, that nature makes great use of it, only producing biodegradable trash. Nature both produces and decomposes with extraordinary efficiency, providing nutrients for fresh growth. It can build for remarkable durability and make the most ingenious repairs.

The efficiency of natural manufacturing can be seen beautifully in the rich pageant of life supported by a single oak tree. How absurd, the old adage: If a tree falls in a forest, and there's no one there to hear it, does it make a sound? Thousands of animal ears are always there to hear, no human presence required. For the Major Oak, one of the oldest and most majestic of its brethren on the planet, standing broken yet unbowed in England's legendary Sherwood Forest, a lack of human presence might well be greatly desired.

The stately sentinel, one of Britain's most revered and most visited, with massive branches spreading a staggering ninety-two feet, has provided essential succor to a vast ecosystem of forest dwellers for over a millennium, including 350 species of insects, a plentitude of birds, innumerable chipmunks and squirrels, raccoons, wild turkey, and deer. Meanwhile the tree has withstood a brutal attack by a fungus, which carved out an enormous cavity in the tree's trunk due to the symbiotic aid of another fungus, a vast and intricate web of mycelium—the finely tendrilled roots of mushrooms that are intertwined with the tree's roots and send it plentiful nutrients. Of course, this tree city of life is fueled entirely by clean solar power, and the tree not only releases no harmful emissions, its massive canopy sucks in vast quantities of carbon from the air. The Major Oak is a closed loop, planet-healthy production system extraordinaire.

A similar efficiency and circularity can be implemented in industrial production. A number of bold pioneers have proven that we really can take an ecological approach to making, selling, and using products of all kinds, and that companies cannot only make good money by doing so, but decidedly outperform competitors who don't.

As with nature's ecosystems, which range in scale from tiny tidal pools to complex localized domains like that of the Major Oak, to the vast stretches of whole forests, oceans, deserts, and prairies, circular economy ecosystems can be created for single products, made by one producer, developed between two companies or clusters of many companies, and, ideally, established for whole cities and regions.

Consider the feat pulled off by business leaders in the small Danish seaport city of Kalundborg. Situated on the northeast coast of the island of Zealand, the city became host to a hub of industrial symbiosis even before the larger scale vision of how to build an industrial ecosystem was laid out by thought leaders. A group of corporate heads developed, step-by-step, the Kalundborg

Eco-Industrial Park. The first symbiotic partnership, inaugurated in 1972, was the building of a pipeline from an oil refinery on the site to transfer excess gas produced by the refining process as power for a nearby factory that makes plasterboard. Over time, several other companies have been looped in, as well as a local farm. One company captures and cleanses the smoke exhaust from a coal-burning electricity plant that then supplies all of the electricity for the park. In the cleansing process, gypsum is produced, which is a component of plasterboard, so that is sold to the plasterboard factory. Two pharmaceutical companies send organic waste from their manufacturing processes to a biogas plant, which uses it to produce natural gas. Water is circulated from company to company, and purified in a closed loop for optimal water efficiency. Tons of waste yeast produced by one of the pharmaceutical companies are converted into food for the pigs at the farm. The biogas company in turn uses animal waste from the farm to produce biogas.

A 2015 analysis showed that since its inception, the park has reduced greenhouse gas emissions the companies would have generated by 635,000 tons. Meanwhile, the companies operate at a lower cost; the same analysis found that they collectively save approximately $26,500,000 (24 million euros) annually by eliminating waste and relying on locally generated renewable energy. The phenomenal success of Kalundborg is inspiring many variations on the theme, both a good deal larger and quite a bit smaller.

Impressive progress in scaling up the model has also been made in South Korea at its Onsan Eco-Industrial Park, which is home to a thousand companies and is the industrial heart of the country. Industrial ecology is also being introduced to the developing world. The World Bank is working with the governments of Turkey and Vietnam, for example, to help them create national-scale systems of eco-parks. Great entrepreneurial creativity is also being brought to developing smaller hubs of symbiosis, showing that even nascent start-ups can avail themselves of the advantages.

One beautiful exemplar is BlueCity, in the Dutch city of Rotterdam, an industrial ecology hub for start-ups. The hub was created by the founders of Rotterzwam, which cultivates mushrooms in coffee grounds collected from area businesses. The carbon dioxide generated by Rotterzwam's mushrooms is used by another food company in the hub, Spireaux, for making a nutrient-rich algae-based paste that's great for veggie burgers. Yet another firm, Fruitleather, makes leather-like textiles out of unsold fruit on the docks of the nearby harbor, which would otherwise go bad. The plan is for the network to continue to grow, evolving further and further toward the zero-waste ideal.

Similarly, in Chicago, the nonprofit organization the Plant has created a complex of food businesses—including a brewery, bakery, coffee roaster, chocolate maker, and several aquaponic farms—housed in a former meat-processing facility in which the waste of one business is used as fuel or ingredients by others, and carbon emissions are captured and sent to the farms to be absorbed by the plants. Symbiosis is also being practiced at a still smaller level, such as by England's Toast brewery, which collects bread that would otherwise be thrown away by local bakeries, restaurants, and grocers, and uses it for making beer.

Valuing Ecosystem Services

As with Herman Daly and so many others, reading *Silent Spring* had a powerful impact on me, opening my eyes to a fundamental flaw in economic structure: companies, by and large, are being allowed to pass the cost of cleaning up the environmental damage inflicted by their modes of production to the public. While the book is best known for its harrowing revelations about DDT and other toxins, Carson also introduced to the public the then-developing concept of an ecosystem, and vividly portrayed the devastation of a range of ecosystems. She bemoaned the evisceration in the West of whole landscapes of wild sagebrush, writing that the "purple wastes of sage, the wild swift antelope and the grouse" had been "a natural system in perfect balance." Way ahead of her time in appreciating that rich ecosystems are hidden away in soil, she devoted a chapter to the havoc being wreaked on soil microbes, explaining that they are vital to soil health and therefore to crop yield. As we've seen, her prediction that continued dousing of farmland with vast quantities of fertilizers and pesticides would lead to devastating soil depletion has come horrifyingly true.

Carson exposed how polluting chemical companies had shunted off onto the public enormous costs of habitat loss and cleanup that they should have been responsible for. They were, in essence, on the public dole. Had they been required to pay the true costs of producing their products, they would have been highly incentivized to innovate in order to avoid the waste and pollution they were creating. In many cases, executives realized that being responsible for those costs would have made their businesses obsolete. Better to spend money on lobbyists to protect their subsidies and block regulations that would require them to cover the cost of their pollution. In the long term, that strategy is rarely effective. Most of the businesses that are the worst polluters eventually fail, either because their business model was never financially viable or because they were eventually forced, via legal action or government regulation, to pay the cost of cleaning up their waste and pollution. Unfortunately, the deception and lobbying dollars usually continue for enough years so that the public can't avoid a significant financial cost and the harm these companies have perpetrated on our environment and health is impossible to fully clean up.

Pioneering circular economists are raising awareness that ecosystems perform a host of valuable services, not only for the creatures that are part of the system but for human society and the stability of the economy. They make the case that companies availing themselves of those services should be made to pay for any of that value they destroy. Better yet, they should be replenishing that "natural capital" they've been depleting. By performing calculations of the money value of services, these pioneers have also helped provide the means of correcting for some of the crude deficiencies of GDP, allowing for the costs of depleted services to be factored into a much more accurate accounting of a nation's economic health. Consider, for example, the enormous hit to the economies of so many coastal communities in the U.S. due to the killing off of fish and shellfish in the Gulf of Mexico by the pollution of oil drillers, refiners, and chemical plants. That has not been factored into the country's GDP, but it has most definitely afflicted the well-being of those communities. Natural capital evaluations are allowing for powerful realizations about the economic as well as environmental damage that taking and wasting reaps.

The concept of natural capital was popularized by British economist E. F. Schumacher, who coined the term in his four-million-copy selling 1973 book *Small Is Beautiful.* Nature's bounty should not be thought of simply as resources, he argued, but as the fundamental capital that makes all of our production and consumption possible. Business leaders understood very well that they had to replenish the financial capital they expend, but they'd been grossly mismanaging natural capital. His opening line minced no words: "One of the most fateful errors of our age is the belief that 'the problem of production' has been solved." How absurd, when "the capital provided by nature . . . is now being used up at an alarming rate"; when "the modern industrial system . . . consumes the very basis on which it has been erected." We need to devise "new methods of production and patterns of consumption" to create a "life-style designed for permanence."

The term "circular economy" came much later, coined in a 1990 academic book by University College London scholar David Pearce, who would be a leader in the field of environmental economics. Pearce, with coauthor R. Kerry Turner, laid out the basic concept of circular versus linear resource use. Also influential was his 1989 book, *Blueprint for a Green Economy,* a major bestseller, which introduced the "polluter pays" principle, recommending that governments charge firms for their pollution. That work is the foundation of the European Union's extended producer responsibility law, proposed in 2020 as part of its Circular Economy Action Plan, to hold companies accountable for the packaging waste they create.

Two other impactful books were Paul Hawken's *The Ecology of Commerce*, and *Natural Capitalism*, by Hawken and coauthors Amory B. Lovins and L. Hunter Lovins (who founded the Rocky Mountain Institute, which researches sustainability solutions and consults for companies around the globe about implementing them). They revived Schumacher's call for natural capital accounting, and Hawken argued that businesses could take the lead in restoring the environment and make good money in the process. A key theme of *The Ecology of Commerce* is that companies that are harming the environment are enjoying an unfair advantage over environmentally conscious firms. "How is it that products which harm and destroy life," he asked, "are sold more cheaply than those that don't?" Because the costs to the environment didn't have to be factored into their prices. Hawken made a masterful case for monetizing natural capital.

The concept is then developed further, setting forth a vision for a natural capitalism system; a capitalist system that not only preserves but rebuilds natural capital. "It turns out," the authors of *Natural Capitalism* asserted, "that changing industrial processes so that they actually replenish and magnify stock of natural capital can prove especially profitable."

Beginning in the 1990s, one of the pioneers of valuing natural capital, ecologist Thomas Graedel, conducted detailed studies of the different types of ecosystems found in nature as models for industrial replication, including none other than that of the oak tree. He also developed elaborate assessments of the environmental impact of natural engineering versus human engineering, such as a comparison of a large suburban house to a beaver dam and lodge. Beavers, not surprisingly, were found to make far superior use of resources. After all, they are widely understood to be, as the Sierra Club writes, the "ultimate ecosystem engineers." As for the environmental impact of a beaver dam, it is far, far greater than that of a human house. In fact, beaver dams are wondrously beneficial overall. They create ecologically rich wetlands, which are vital life sources for thousands of species. The ponds they form trap copious amounts of carbon in the sediment that drifts down to the bottom. They also raise the water level of streams, so much so that they often create new tributaries, as well as purify streams of toxins and infuse them with natural fertilizers, thereby rejuvenating depleted fish populations. So keen are environmental stewards to reintroduce beavers into terrain where they went extinct that a number of them were actually parachuted into remote lands in Idaho, to great success (though one imagines a good deal of unnecessary trauma for the beavers). In England, where beavers went entirely extinct in the sixteenth century, several reintroductions are under way.

Beavers are tremendous builders of ecosystems that generate multiple benefits. Consider just their restoration of wetlands, which are often called nature's kidneys because they filter out toxins from water and balance levels of natural chemicals like nitrogen. With communities all around the globe already facing critical shortages of fresh water, massive investments in water filtration and conservation have been made, but the water crisis will be getting a good deal worse. One analysis concludes that by 2030 the global demand for fresh water will exceed its supply by 40 percent.

Placing the appropriate monetary value on nature's resources and services is a powerful lever for inspiring—and for governments requiring—much more intelligent use of them. Despite the extraordinary complexity of the task, a number of ingenious means of doing the job have been devised by individuals and organizations around the globe. They have taken the concept of natural capital and made it a practical component of business planning, which is helping to make a highly persuasive case for holding polluters more accountable. They've also worked out schemes whereby companies and communities that help preserve and restore ecosystem services are being compensated for what they do.

David Pearce was a leader in devising specific calculations of ecosystem services values. He was a masterful number cruncher. One calculation: the value of carbon sequestration provided by forests might be as high as two thousand dollars per hectare (about 2.5 acres). To give a sense of how that translates into large-scale forest valuation, consider that there are 32.5 million acres of forest in New England, which would amount to $26 billion in services annually. Another valuable service performed by forests is water table replenishment. Calculations like those Pearce worked out have been used to create successful programs in Mexico and Peru that compensate indigenous communities for the work they do to preserve the forests they live in and near.

A growing legion of environmental economists have followed Pearce's example and provided many compelling eco-services valuations, helping to build the case that the costs should be paid by the companies whose operations have contributed to degradation. For example, one recent estimate of the annual value of pollination by bees and butterflies to food production globally calculated the dollar figure at $577 billion—none of which is being paid for by the corporate agricultural giants whose profits depend on it, and who are contributing so much, with their pesticides, to horrifying die-offs of butterflies and bees all around the world. The cost of the decimation of pollinators is subsequently being paid by many farmers, who must dish out rental fees to bee cultivators to make use of their colonies. And inevitably the costs of so much other environmental damage are being paid by the taxpayers, whether for cleanup operations, water filtration and desalination plants, fishery restocking, and so many other conservation and restoration measures.

A newly developed practice, true cost accounting, calculates how goods and services would be priced if the full cost of resources and eco-services, and of the ecological and human harm done by their manufacture, sale, and disposal, were factored in.

This has helped generate buy-in to the argument in the business community, with the help of major champions like Seventh Generation founder Jeffrey Hollender. He became a convert when he learned why Seventh Generation's now ubiquitous bath tissue, made from 100 percent recycled fiber, had to be priced higher than that of competing brands. Those competitors were benefiting from tax breaks of about $1 billion a year on the sale of virgin timber; effectively receiving a massive subsidy from the American public, without its knowledge, for buying paper pulp from felled trees. "That's when I knew the game was fixed," he recalls. Not only were the competitors not paying true market price for raw material; they weren't paying their share of the costs of forest-services depletion either.

The true costs of a wide range of common household purchases have now been calculated by a number of organizations, such as the British consultancy Trucost and Dutch social enterprise True Price. According to one analysis by True Price, the true cost of a 250-gram (approximately half a pound) bag of conventionally grown coffee would be $5.17, compared to a cost of $4.58 for the same size bag of sustainably grown coffee. Rather than considering the higher costs we so often have to pay for sustainably made products as an unfortunate but necessary premium, when seen through the lens of true costing, suddenly we can appreciate that the public at large has been paying at least as much as that premium for unsustainable goods—just not in the sticker price.

True cost accounting has become a powerful tool with which companies can assess their environmental impact and identify ways they can redress damage. One corporate leader who seized the opportunity is Jochen Zeitz, the former CEO of Germany-based Puma and the current chairman and CEO of Harley-Davidson. He collaborated with Trucost to create the first environmental profit and loss statement for a company, released publicly in 2011, which indicated that the total cost of Puma's environmental impact for 2010 was 145 million euros. Zeitz tells me that the EP&L was instrumental in advancing Puma's sustainability efforts. We'll hear from him later about what a boost going more circular was for Puma's business, but for now, suffice it to say that during Zeitz's tenure, the company's stock value rose 4,000 percent.

Selling Performance in Place of Waste

Walter Stahel has been particularly persuasive, emphasizing that circular production is superior not only because it is sustainable, but also because it's higher performance.

Still sprightly at seventy-three, Stahel entrances audiences at his many public talks and on YouTube videos with the intensity of his devotion to designing products for longevity, reuse, and repair. No mere proselytizer, he has practiced what he preaches in his own life; he happily recalls that when he had the body of his thirty-year-old 1969-model Toyota entirely remanufactured, the first day he pulled it into his driveway a neighbor called out to him, "I am so glad you finally bought a new car!" Reflecting on the comment, Stahel laments, "Quality is still associated with newness, not with caring; long-term use as undesirable, not resourceful."

In the mid-1970s, when Stahel was working at the innovative Battelle Research Centre in Geneva as the head of product research, he was commissioned by the European Union to explore the possibilities for using less energy in manufacturing, prompted by the recent 400 percent spike in oil prices. Stahel's groundbreaking report argued that product-life extension, through refurbishment, was key not only to less energy use, but to the creation of good jobs, replacing human power for energy. He cofounded the Product-Life Institute to research possibilities for cradle to cradle design and closed-loop manufacturing, product delivery, and recovery, and to consult with companies about implementing them.

He has hammered home the fact that by making more efficient use of natural resources in the initial production process and then extending their use as products for as long as possible rather than recycling them fairly quickly, we get much more performance bang for our resource buck. This was the central premise of what he dubbed a "performance economy" in a book of that title, though he now prefers the term "circular economy."

Stahel has vigorously championed selling a product's performance rather than the product itself, now called product-as-a-service, one of the most effective models for incentivizing building to last and investment in repair and recovery. The manufacturer retains ownership, charging only for use, and therefore also retains product responsibility. Companies have every reason to design for longevity and market reuse, and to invest in repair, refurbishment, and upgrading. While product-as-a-service has come to be regarded as a post-industrial digital economy innovation, some surprising industrial-era stalwarts (in some of the highest carbon emitting sectors, no less) have been in the vanguard in developing models.

Rolls-Royce, a leader in both car making and jet engine manufacturing, began selling engine time in lieu of engines back in 1962, with its Power-by-the-Hour program for its Viper business jet engine. It's expanded the program to include engines for large commercial planes, and prodded rivals General Electric and Pratt & Whitney to do the same. Tire maker Michelin has been selling tire service by the mile to trucking fleets for twenty years. The company monitors tire use through digital tracking and retrieves tires for repair, restoring, and regrooving them, through its own fleet of mobile workshops. A particularly gratifying more recent example, given the role of lighting giant Philips in the Phoebus lightbulb cartel, is the company's "pay per lux" program, inaugurated in 2015. Philips retains ownership of lighting systems, bulbs and all, installing LEDs and performing constant remote digital monitoring for maintenance as well as for optimal lighting levels throughout the day and night, cutting down significantly on electricity use by companies. In one heartening case, Amsterdam's Schiphol Airport, which burns the amount of electricity a day equivalent to that used by fifty thousand houses, contracted with Philips and reduced its lighting expenses by 50 percent. Their successes provide proof, from long-term experience, of the value created for the environment, shareholders, and customers.

In a robust closed-loop production economy, Stahel has stressed, recycling should be a last resort, held off for as long as possible. Though a remanufactured car engine might be worth $4,900, for example, as scrap metal it plummets to $160. What's more, restoring engines would not only significantly reduce the environmental devastation of extracting raw metals (more on that later) and the greenhouse-gas emissions from superheating them, it would also create lots of high-skilled jobs.

When it comes to the end of product life, recycling, Stahel has also been a leader, championing design for disassembly and materials recovery. He has urged that products be crafted so that all the materials they're made from are easily separated again. With too many products, materials are blended in ways that make them veritably impossible to prize apart. Consider the typical tube of toothpaste. For optimal squeezability, they're made from a mix of plastic and aluminum, which are so devilishly difficult to separate that the tubes are for practical purposes unrecyclable. What a problem crying out for a solution, given that by one count 400 million toothpaste tubes in the U.S., and 1.5 billion globally, are trashed annually. Thankfully, Colgate has stepped up to the challenge, designing a tube made of recycled plastic, sans aluminum, with just the right squishiness. The feat wasn't easy; it took five years. Product design hurdles, like engineering squishiness, should never be underestimated.

Business innovators have been making so many great strides in developing products and business models that fit Stahel's bill of reduce, reuse, remake, recover, and renew. Cradle-to-cradle product design guidelines spearheaded by William McDonough and Michael Braungart and others have provided much-needed inspiration for designers to tackle vexing challenges. Since founding the Cradle to Cradle Products Innovation Institute to promote specific standards and provide C2C product assessment and coveted certification, the team have been marvelously influential in stoking public demand.

When it comes to reducing waste, an especially gratifying example to me, given my run-in with foam maker Dart, is Green Home, the South African producer of entirely plant-based, wholly biodegradable food packaging. While Dart dithered in developing foam alternatives, in a single year Green Home founder Catherine Morris went from having no knowledge whatsoever about making packaging of any sort to working out how to make replacements for plastic out of a trademark brew of sugarcane waste, wood fiber, wood cellulose, bamboo, and plant-based starch. A video producer inspired by seeing biodegradable food packaging on a trip to Thailand, Morris is a sterling exemplar that where there's a will, there's a way. She's just one of a legion of innovators, both building start-ups and working in corporate labs, who are driving explosive growth in the biodegradable packaging business.

A runaway success with product reuse has been Patagonia's Worn Wear business line, selling used Patagonia items both in the company's stores and online. Because Patagonia clothing is built to last, customers are able to bring trade-in items in good shape for store credit; the items are washed and sold, looking brand new. The popularity of the prototype program surprised the company. Senior director of corporate development Phil Graves says that it started as "a cool idea to keep our gear in use longer, but now it's this fledgling e-commerce business that we want to grow in a big way. The goal is to encourage every major brand to have their own recommerce site behind their apparel." They're well on the way, with The North Face, Macy's, J.Crew, Burberry, and many others getting in on the game. As we'll see later, this innovation has the potential to take an enormous bite out of greenhouse-gas emissions.

In the design-for-recovery-and-repurposing category, whole buildings are being constructed with eventual disassembly and reuse in mind. An inspiring case in point is the city hall in Venlo, the Netherlands, which was designed with construction and deconstruction blueprints, according to cradle-to-cradle guidelines developed by the C2C Expolab, a consultancy that works with property developers. One extraordinary innovation incorporated in the building is self-gripping bricks, which are held together with metal fasteners rather than mortar, allowing them to be easily unfastened from one another. Given that the demolition of buildings accounts for 90 percent of the waste created by the built environment, which in turn accounts for as much as two thirds of all solid waste, the potential of this area of innovation is staggering.

Regarding designing for renewal, one of the first and most appealing accomplishments was the construction of the visually stunning visitor center at the VanDusen Botanical Garden in Vancouver. The building embodies the principles put forth by Californian landscape architect John T. Lyle, the pioneer in the field of regenerative design, another of the streams of thought feeding into circularity. Lyle championed the concept of "living buildings," which are intimately merged with their surrounding habitat and actually replenish its ecosystem.

The VanDusen visitor center does so brilliantly. Its gracefully curved earthen walls, undulating grass-covered roof "petals," and entirely closed-loop solar-powered energy and water systems were inspired by the beauty and extraordinary ecological ingenuity of wild orchids. The lush grass grown on its multitiered roofs, which gently slope down into the surrounding natural landscape, was designed to provide an enticing "salad bowl" for the squirrels, butterflies, rabbits, and other critters that happily frolic over them. Rainwater is captured for all water use, which is filtered and released to replenish the local groundwater. The building is a stirring testament to the potential for us humans to live *with* nature rather than separated from it, even in the midst of urban metropolises. As we'll explore later, many ingenious variations on the theme have now been constructed globally.

While today the most transformative examples of circularity are developing with individual products and in individual companies or buildings, larger scale change at the city, regional, and national level is also advancing. Asian economies are enthusiastically adopting industrial ecology, and the European Union's proposed Circular Economy Action Plan, aimed at ensuring that producers are responsible for any costs related to the disposal of their product in a landfill, is another major development.

While for start-ups going circular can be an integral component of the business from the get-go, it still involves development in stages. For established companies, it requires focusing initially on particular products or product lines, and most often begins with the simplest processes—switching to renewable energy, and replacing virgin materials with recyclables or ecologically healthier ones, like sustainably grown wood and cotton. Creating new product designs with modular components for easy repair or recycling and developing reverse supply chains to take back and recirculate their products are more challenging stages. But as we've seen, even blue-chip corporate behemoths are pushing forward on these fronts.

A Matter of Consciousness

The wind was up to 30 knots by now, a rise of 5, which was pushing us closer and closer to the side of the berg; though we made it safely round, we passed it well within a mile. As we rounded its northernmost point, I was stunned to see two enormous ice caves, perfectly arch-like in shape and sinking away to darkness deep within the berg itself. Each was large enough to sail the Kingfisher into—a trip, though, from which there would be no coming back. Its sides were tinted aqua-blue and it looked so white against the inky-grey sea and darkening sky. There was an aura around this berg, a feeling of complete and utter isolation. I knew that it was highly unlikely that human eyes had ever seen it before.

So wrote the Englishwoman Dame Ellen MacArthur in her account of her solo navigation of Earth on the 60-foot *Kingfisher*, making her at twenty-four the youngest sailor ever to complete the grueling Vendée Globe race. "We" refers to her, the boat, and the legions of supporters she felt were with her, emailing and texting her a steady stream of good wishes. Though she came in second in that challenge, she went on five years later, in 2005, to break the record for fastest solo circumnavigation, for which she became the youngest person in the modern era to be knighted, at twenty-eight.

MacArthur says her globe-encircling journeys led her to a profound realization. "When you set off around the world," she says, "you take with you everything you need for survival. . . . In the Southern Ocean, you're 2,500 miles away from the nearest town." This was her own Spaceship Earth eye-opener, and with all the vigor with which she went about sailing, she has since promoted the advancement of circularity, founding the Ellen MacArthur Foundation in the service of the cause. In 2012, the foundation collaborated with McKinsey to produce a galvanizing report laying out a detailed and highly persuasive description of how firms of all sorts can embrace circular processes. The foundation has gone on to sponsor a great deal of research and many more influential reports that have been read widely in the government and business worlds and helped catalyze the current innovation and financing boom.

Ellen MacArthur wrote one night in an email to supporters during the Vendée race, "As I finish this message something catches my attention through the window above. The moon is beaming its presence through the tiny openings in the clouds . . . round and beautiful . . . a gentle reminder that it's that very same moon you can all see in your hours of darkness."

It's been just over fifty years since Apollo 8 raised consciousness about how closely interconnected the lives of all of us humans on Earth are with the health of the planet. It's time now to raise consciousness again, about how capable we are of ensuring our planet remains habitable and supportive by harnessing the wisdom of its superefficient and regenerative circularity.

PART
TWO

A
Wealth
of
Circular
Solutions

For the Love of Forests

S TANDING ON THE LOADING DOCK of the Pratt Industries factory on the edge of the Arthur Kill waterway on Staten Island's western shore, after my walk through Freshkills Park, I watch a barge heaped high with 400 tons of New York City's daily paper collected for recycling glide silently in for mooring. With the Manhattan skyline shimmering in the distance, I recall reading a comment CEO Anthony Pratt made to a reporter shortly after this facility opened in 1997. Looking over to the city, he told her, "You see a city. But if you're interested in trash, it's really an urban forest that renews itself every day. The potential here is enormous."

I learned about the remarkable Pratt family when I worked in the Bloomberg administration, and the saga of how Anthony's grandfather went from being a nearly penniless Jewish refugee from Nazi Europe to an Australian box-making and recycling magnate. Pratt Industries is a standout example of the extraordinary opportunity for entrepreneurship in building the circular economy. One of the world's leading makers of boxes, Pratt has developed an integrated circular production system, and the Australia-based firm brought its innovative closed-loop approach to box manufacturing to the U.S. in the 1990s. Its success has led to the development of a hundred facilities in twenty-six states that use 100 percent recycled paper. Their control of their own supply chain maximizes efficiency and reduces volatility. In most cases, their recycling facility, mill, and box-manufacturing plant are colocated; the company owns paper recycling facilities that have long-term municipal contracts, which feed their pulp mills, which in turn feed their box factories. At its New York City facility, one of the largest in the world, an especially efficient closed loop is achieved by devoting much of the cardboard produced to making pizza boxes for the pizza-crazed city. Paper mushed and reconstituted in the morning can be on its way to pizzeria counters in just twelve hours.

Pratt's circular integration is a great example of the advantages of local closed-loop production that Walter Stahel highlights. Not only has Pratt cut out middlemen, it has reduced its energy consumption and emissions by limiting its transportation needs. In short, the company has created highly efficient eco-industrial complexes that continually invest in state-of-the-art sorting and recycling equipment. It's that drive to reduce waste and cost in their supply chain that's taken the company from a tiny operation in a factory the size of a typical living room to a $3-billion-plus per year revenue-generating powerhouse, still owned and led by the Pratt family.

As a biography about Anthony's father by James Kirby and Rod Myer recounts, box making wasn't at all on company founder Leon Pratt's mind when he fled the Polish city called Danzig, now Gdansk, on the coast of the Baltic Sea. Awarded to the Poles in the breakup of the Prussian Empire following World War I, the city was, however, populated mostly by people who identified as German, and with Hitler's rise, a local Nazi party formed and won control of the city's government in elections in 1933. Persecution quickly followed, and many of Leon's Jewish friends fled. But he and his wife, Patricia, were reluctant to leave. Leon owned a small bike shop, and Patricia gave birth to a son, Richard, in 1934. They had hoped to build a comfortable life there. But as the Nazi invasion loomed in 1938, they decided they had no choice. Doors were being closed to Jews all around the world, but an exception was Australia, and that, they decided, was where they would settle.

They arrived in Melbourne with only two thousand British pounds and no prospect of work for Leon. Improbably, he decided to become a fruit grower, despite having no experience whatsoever in fruit farming, using all his money to purchase sixty acres in the soil-rich region around the city of Shepparton, which is known as Australia's food bowl. Tackling novel challenges seems to be a family talent: Leon made a go of selling fruit, but the boxes available for packing fruit were shabby, he thought. Why not build a better box?

He and a few relatives who had followed him to Australia joined forces with an engineer friend and cobbled together a box-making factory, or an approximation of one. Their first machine was concocted—aptly enough for a company that would become a recycling pioneer—out of scrap parts by two engineers for hire who knew nothing about how to make one. They opened their first factory in a space just fifteen by twenty feet.

Anthony's father, Richard, took over in 1969 and turned what was still a modest operation into one of Australia's largest companies. Its competitive advantage from early on was that 100 percent of the pulp used is from recycled boxes. Seeing enormous opportunity in America's failure to build its paper recycling capacity, Anthony built up Pratt's American operations.

Anthony cuts a striking figure, with once bright-red hair now toning down to a golden brown. He's always ready with a broad smile and clever quip, and he doesn't shy away from the media. A big fan of boxer Muhammad Ali, who became a good friend, he remarked to one reporter that he likes to think of Pratt Industries as the world's second greatest boxer. When it comes to business, though, Anthony couldn't be more serious; he's a relentless innovator. But the company's advance into the U.S. wasn't treated with much respect at first. Even as Anthony almost immediately began making an impressive success of the American business, the company was viewed by much of the industry, he says, as "a schlock recycler." When a number of corporate giants and major cities, including Walmart and New York City, came calling in pursuit of a partner to help achieve their sustainability goals, Pratt suddenly wasn't looking the slightest bit schlocky anymore.

Watching the barge pull gracefully up to the Pratt dock, gently nudged forward by a classic little red tugboat, I'm impressed that even the transport of the paper to the facility is done by optimizing for energy efficiency. To complement the barge transportation, the Bloomberg administration approved the reconstruction of a mile of tracks to the Pratt facility so paper could be brought by more energy-efficient rail instead of gas-guzzling trucks, as part of the city's climate change remediation master plan.

Pratt Industries and others leading the way in circular papermaking are vital aides in furthering one of the most potent means of pulling greenhouse emissions back out of the air—saving and restoring natural forests. They are one of the planet's most voracious consumers of carbon dioxide, storing vast quantities of it in their trunks and root systems. As we'll see in a bit, great innovation in papermaking and remaking is one of the most impressive fronts in the advance of the circular economy. One pesky holdout, however, has been paper cups. They seem such a simple creation, but that's a misconception, as we at Closed Loop Partners learned in our NextGen Cup Challenge.

Innovating a New Way to Drink

As my team at Closed Loop Partners began searching for start-ups to back, we began to see that there was a need for an innovation incubator to tackle challenges that didn't have investment grade solutions. So we decided to take on that role. I created our Center for the Circular Economy (CCE) innovation accelerator to conduct research into how to solve such puzzles, and to bring together players to do so in collaborative efforts. We've been able to form partnerships with many leading consumer goods brands and retailers, such as Nestlé and Walmart, as well as leading NGOs, like the World Wildlife Fund, and some of the world's most respected specialists in design innovation, including renowned design firm IDEO. One of our main activities is managing competitions that challenge entrepreneurs to develop solutions to specific problems, providing funding support and expert business-building guidance to winners. Another focus is on precompetitive collaborations among major brands who share challenges in order to provide solutions at the scale required to get funding.

I hired Kate Daly, another former member of the Bloomberg administration, to manage CCE. One of the first initiatives she developed was the NextGen Consortium, with founding partners Starbucks and McDonald's. The consortium aims to advance recoverable solutions for the fiber, hot and cold, to-go cup system, exploring recyclable, compostable, and reusable cup designs. Currently with most cups there are two issues. First, they are often contaminated with food or liquid. Second, paper cups have a thin plastic liner to avoid leakage, meaning that the cup has little to no value in the recycled paper market. The result is that far too many of the estimated 250 billion paper cups produced globally every year end up in landfills.

Both Starbucks and McDonald's had been working for years to find a fully sustainable alternative to paper cups, and they partnered with us to lead a consortium of retailers and brands also interested in a solution. The World Wildlife Fund and IDEO joined as advisers.

In a clear sign of just how vibrant the community of circular innovators is, we received 480 cup design submissions from around the world, ranging from start-ups to established companies. A panel of twelve experts in innovation, sustainable packaging, and investing helped us whittle the submissions down to twelve finalists that are already in business with their solutions or have some good proof of concept, and we've provided them with funding and access to our accelerator service.

We also held a pitch competition, in which the companies presented to a panel of judges. As the pitches got under way, the frisson among media, industry professionals, and potential funders was electric. Any solution chosen could land a contract to service many of the world's largest restaurant brands that were partners in the NextGen Consortium.

Up onto the stage first was Ayca Dundar, founder of SoluBlue, the England-based innovator of a beautiful transparent Caribbean-blue cup. It looks like plastic but is made of 100 percent plant-based material that the company reports is both biodegradable and safe for marine animals to consume. SoluBlue aims to sell them to businesses and then take care of the biodegrading for an additional fee. "You don't just pay for our cups," Dundar explained. "You pay to make them go away." Clearly relieved to be finished with her presentation, she immediately scooted offstage, but quickly returned to field questions from investors. "How do you get to producing billions of cups," asked Abe Minkara, then managing director at Mark Cuban Companies, "and why do you think companies will pay you to decompose them?" SoluBlue has been advised by an innovation consultancy to start small, Dundar responded, but the technology can scale well, and companies already pay for garbage removal. Good answers.

Up next was Fabian Eckert, cofounder of RECUP GmbH, a German purveyor of appealing pastel green reusable cups-as-a-service, with the cups made from recyclable plastic. He aims, he said, to launch "a coffee-to-go revolution." "Renters" can take them to go and stroll around town with them, then return them to any one of a number of participating shops—all return locations are easily identified through a mobile app—getting back an initial one-euro deposit, after which the cups are given a sterilizing wash. After two years, the company is operating in over three thousand shops throughout Germany. What's more, they plan to embed the cups with a tiny computer chip so they can track them and optimize recovery. Another great example of the high-tech return of returnables, I thought. But can that really work, beyond a niche of especially dedicated green-conscious consumers, I wondered? How many people will be willing to pay the equivalent of a $1 deposit and walk around to find a drop-off spot? It seems to be working in Germany, but what about in the U.S. or China? That's precisely the kind of question the NextGen Consortium is seeking to answer.

Most of the finalists have pursued simpler solutions, creating biodegradable and recyclable paper cups by inventing alternative inner coatings that don't devalue the paper cup in the paper recycling market. One such cup, developed by U.S. start-up Footprint, defends against both water and oil.

Innovations almost never catch on at large scale right away. Consider the case of the paper cup itself. It was invented in 1907 by Boston lawyer Lawrence Luellen, who was inspired by new scientific research showing that the scourges of cholera, tuberculosis, and diphtheria, which afflicted so many at the time, were spread

"SPARE THE LITTLE CHILDREN!"

From the Kansas City Post.

largely by people sharing the glass and metal "common cups" that were attached to public drinking fountains and coolers.

Luellen dubbed his solution the Health Kup and founded the Individual Drinking Cup Company. He invented a vending machine, to be installed next to water fountains, that dispensed a single small cup—holding only about two good gulps—purchased for a penny. Luellen introduced his wonder to the market in 1910, but despite a massive public education campaign against common cups, featuring macabre illustrations and screaming warnings like "Death in School Drinking Cups," much of the public strenuously objected to paying for the cups.

To force their hands, states began passing ordinances banning "the promiscuous use of common drinking cups," as a 1911 Texas law put it. Objections were at first fierce. "The annoyances of asinine legislation were realized to a marked extent in New York during the past week," wrote one reporter, "when the law against the public drinking cup was enforced."

As common cups steadily disappeared and more and more water-vending machines cropped up, from train stations and bus depots to the elegant flagship Lord & Taylor store on New York's ritzy Fifth Avenue, many people started defiantly carrying metal cups with them, made in ingenious collapsible styles. Some came in gorgeous leather traveling cases, examples of which can be found for sale on eBay. Though plenty of models are still in production today, the assiduously instilled throwaway mentality has made them little used.

As for paper cups, widespread embrace of them was only achieved in 1918, due to the horror of the 1918 flu pandemic that killed an estimated 675,000 Americans and 50 million people worldwide. Our experience with the coronavirus and the rapid embrace of face masks, by most people, gives us a sense of the urgency. Luellen suddenly became a hero.

Developing new products is often deeply perplexing, and lots of experimentation is required, almost always with plenty of false starts and grueling hours of working details out. If reinventing the paper cup weren't so tricky, both McDonald's and Starbucks would have found solutions on their own.

The biggest surprise of the competition for us at Closed Loop Partners was how complex the problem of paper cup use is. Cups have to meet many stringent standards. They have to hold liquids at very high temperatures, and yet they've got to be comfortably held. Lids have to fit on them securely, to avoid any leakage. Kate worked with partners from IDEO, Starbucks, and McDonald's to assess all the performance and safety challenges, and we consulted with recyclers about the tricky logistics of recovery and repulpability. But even with all such issues attended to, how consumers would respond to the designs was, as ever with innovation, a complete unknown. The NextGen Consortium included tests out in the wild—in coffee shops around San Francisco—of a select set of the finalists. With results from that testing, NextGen is heading into the phase of assisting brands in potentially adopting solutions and scaling them.

Recently, two major paper mills agreed to accept recycled paper cups as part of their feedstock, after two years of partnership and collaboration with the NextGen Consortium. Replacing difficult to recycle cups will be a years-long process, but the payoff will far outweigh the effort. As Colleen Chapman, vice president of Global Social Impact for Starbucks, says, "This is a moon shot for sustainability."

What's been especially gratifying is the cooperation among world-class brands, and two major competitors no less. We've also seen this in another initiative to boost plastic recycling I'll describe later, which is bringing together fierce industry-leading competitors PepsiCo and the Coca-Cola Company in precompetitive problem solving. That world-leading innovation specialist IDEO participated is another definitive sign that we've reached a circularity innovation tipping point. Many of the most innovative minds of our time are focusing on developing products and services that utilize circularity as their core strategy.

We don't know yet know which cup solution will win, or if

there will be multiple solutions for both recyclability and reuse. Every tree saved by cup recycling or reuse will be significant in the climate battle. Of course, recycling of all paper products is vital. One of the investments that must urgently be made in advancing circular solutions is in state-of-the-art paper recycling facilities; the amount of paper we consume, especially with the box boom in e-commerce, isn't going to dramatically decrease any time soon.

The Great Paper Boom

Shortly after twenty-nine-year-old William Randolph Hearst assumed ownership of the money-losing San Francisco newspaper *Daily Examiner* in early March 1887, he transformed the paper into a sensation by plastering its front page with dramatic plot summary–like headlines, in huge bold type, decrying dastardly crimes and criminals. Consider this mini penny dreadful from March 31 headlining the attempted killing of Russian czar Alexander: ASSASSINS. UNEASY LIES THE HEAD THAT WEARS THE CROWN OF ALL THE RUSSIANS. HUMAN BLOODHOUNDS ARE HUNTING HIM TO DEATH. PERIL IN EVERY CONCEIVABLE FORM HAUNTS HIS PATH. DEATH LURKS IN EVERY CORNER, DOGS EVERY STEP (and that's an abbreviated version). When a fire destroyed the gracious old Hotel Del Monte, south of San Francisco, Hearst devoted the whole first page to the story, headlined HUNGRY, FRANTIC FLAMES. LEAPING HIGHER, HIGHER, HIGHER WITH DESPERATE DESIRE AND RUSHING IN UPON THE TREMBLING GUESTS WITH SAVAGE FURY. Suicide attempts were salaciously exploited, as the front-page story on Christmas Day of that year proclaimed: SANTA ROSA SENSATION. REPORTED ATTEMPTED SUICIDE OF MRS. MARTIN, A DOSE OF POISON, KEPT ALIVE BY ELECTRICITY, which claimed Mrs. Martin was being "kept alive by the use of an electric battery." Just how the battery was involved, the writer doesn't say. One would hardly think the story good Christmas morning cheer, but in fact suicides and attempts proved especially popular fodder. Hearst knew his readers.

He had quickly perfected the art of newspaper melodrama,

learning from its master craftsman, Joseph Pulitzer, publisher of New York City's *The World*. When Hearst bought the all-but-defunct rag *New York Journal* to take the city by storm, he and Pulitzer went to battle in what was to become known as the yellow journalism newspaper war, or, as one English journalist described it at the time, "a contest of madmen for the primacy of the sewer."

Their shameless sensationalism turned newspaper reading into a mass-market affair, driving an enormous increase in newsprint consumption. The volume of newspapers printed in the U.S. doubled between 1880 and 1890, and then again by 1900. Paper production was off like a rocket, and it's continued to grow ever since.

In 1975, *Bloomberg BusinessWeek* published an article titled "The Office of the Future," in which Xerox CEO George E. Pake predicted that computers "will change the office like the jet plane revolutionized travel." He was right in many respects, but ironically for a copier company chief, way off in expecting screens to eliminate paper. While it's not in the least surprising that the ease of xeroxing propelled office paper use, what was a surprise was that the confluence of great decreases in the costs of home printers and the emergence of the World Wide Web drove paper use up further still. A 1999 report estimated that at that time, North Americans were consuming 11,916 sheets of paper annually per capita and Europeans 7,280; annual worldwide paper consumption grew from about 250 million tons in 1990, the year after the web first went live, to 400 million tons by 2010. Overall paper use has increased 126 percent in the U.S. over the past twenty years.

With ever more information so effortlessly at our fingertips, it turned out we wanted to put it in print. Why? There's no definitive answer, but the continued popularity of print books following the advent of ebooks suggests it may, in part, be due to the psychology of reading. Many of us really like the feel of a paper page, with good reason; studies have shown that reading comprehension is somewhat higher for reading in print versus onscreen.

The long-anticipated decline in office paper did finally begin in 2008, though by only about 2 to 3 percent a year. But in the developing world, its consumption is increasing, along with the increase in sales of office and home printers.

The current demand for paper comes at the cost of an estimated 4 billion trees felled every year around the planet. That's a difficult number to grasp, so consider these eye-opening stats: the paper for printing a single Sunday run of *The New York Times* reportedly requires seventy-five thousand trees, and if all newspaper readers in the country recycled just one tenth of the papers they read every year, 25 million trees would be saved.

The boom in e-commerce has driven annual growth rates of the global box market to about 4 percent a year, and that rate is expected to hold for the next few years. A 4 percent annual growth rate means demand for e-commerce boxes will grow by 50 percent over the next seven years. The boost in e-commerce has not so much caused a spike in box use as it has shifted it to direct-to-consumer uses of boxes and away from shipping to retailers. This has brought the average box size down somewhat, as those coming to our homes are generally considerably smaller. Still, we're using a whole lot of boxes. The good news here is that, as paper scientist Gary Scott (no relation to the Scott Brand empire) told me, "the paper industry has become quite green."

Paper mills once belched sulfurous billows out of their smokestacks, and anyone who lived near one, or even drove by one with the windows down, will never forget the smell. Gary Scott grew up about a half mile from a mill in Wisconsin and recalls that rainstorms generally swept in from the direction of the mill, so whenever the stench of sulfur wafted into his neighborhood, they knew rain was coming. His father and brothers worked at the mill, and so did he all through high school and college, taking night shifts. When I asked why the mill stayed running through the night, he explained that the equipment was so complex that it took as long as twelve hours to get up and running, so shutting down made no sense. Though the process of making paper is, in its basic steps, quite simple, it's come a long way from the days of dissolving rags.

The invention of wood pulping in the 1840s made it possible to mass produce paper, which was for decades done mechanically by pummeling wood chips into a powder (hence the phrase "beaten to a pulp"). Virgin newsprint is still made from mechanical pulp, but for most higher-grade papers, mills have switched to chemical pulping—now for the most part using nontoxic chemicals. Scott says the billows disgorged by mills today are almost entirely steam.

Innovation is making paper recycling more circular. Have you ever wondered whether envelopes with cellophane windows, which so many bills come in, are recyclable? What about flyers with special offer cards glued onto them? Wouldn't those windows, staples, and glue need to be removed? Yes, and the best recycling machinery has been ingeniously devised to do so. It can also separate the plastic coating from paper cups using a series of screens, centrifuges, and a nifty little trick with air bubbles. Anyone who's thought the business of recycling is really just glorified junk removal would become an instant convert watching the process unfold. The sophistication and magnitude of operations at leading facilities is a beauty to behold.

At one plant, the Resolute Forest Products recycling mill, snug up against the Monongahela River in West Virginia, tons of mixed-paper bales full of orange juice and milk containers, frozen-food boxes, envelopes, and home and office printer paper are first sprayed with a concoction that identifies lower-grade paper, which is pulled out for making a different brew. The clear spray turns colors on a spectrum from red to purple, in a chemical reaction akin to the way those pH sticks we all tested liquids with in biology class turned red when dipped in acid. The remaining tons of mixed paper are then sent by a conveyor belt, which uses radiation-beaming sensors to determine their precise composition and weight, into a massive liquifying machine called a drum, filled with exactly the right amount of paper to get a perfect consistency of pulp. This resulting gray slush is then sent through a filter that catches whatever staples, paper clips, or bits of broken glass might be lurking.

Next, it's off to a series of centrifuges for a vigorous spinning that pushes heavier nonfiber components like foil or plastic coatings out to the edge while the lighter paper fibers gather in the middle and are sucked out the top. The fibrous mush that is created looks like thick papier-mâché and is then sent for kneading. Machines with jagged steel teeth pummel the pulp to dislodge ink particles from the fibers, in the process called "deinking" (which can be done in many other, more sensitive ways, such as using organic enzymes to eat the ink off). With the pummel method, the doughy pulp is sent to a flotation tank, in which it's rewatered, turning it again into a fine slush. Air bubbles are then pumped into the tank and the ink particles attach to them and float to the top—forming a heady foam like the top of a glass of beer—which is skimmed off. But the pulp isn't quite finished yet. It's kneaded again, this time to dislodge glue particles, which, in another flotation tank, hitch their own air bubble ride out.

The Resolute plant is world-class in its efficiency, sustainability, and mixed-paper processing capability, which is why it was acquired in 2018 by the U.S. subsidiary of the Chinese paper goliath Nine Dragons. The purchase underscores that while the media in the U.S. have been so quick to pronounce recycling doom, the Chinese have focused on its auspicious future. As for workers in the economically distressed coal-mining mecca of West Virginia, paper recycling is proving that the future of jobs, as well as the planet, is in circularity.

Recycling plants must be upgraded to Resolute's standard. But limiting the destruction of trees through recycling is only one crucial forest preservation solution.

Circular Forestry

Paper companies have been key in supporting the adoption of regenerative forestry methods developed by scientists who study forest growth and natural destruction and renewal. These scientists have ventured into teeming, thick canopies of rain forests and the deepest darks of rare old-growth woods to learn how to plant and groom forests in a respectful, regenerating manner—

following the lead of nature's own circular systems, allowing forests to flourish while also providing a wealth of wood and food resources. The international nonprofit Forest Stewardship Council (FSC) monitors forest management practices, and many paper companies and consumer goods makers have restricted their supply to FSC certified forests (and display the certification on their packaging). When managed according to best practices, these tree plantations, which provide a large portion of wood for paper pulping and lumber, are sophisticated, ecologically sensitive operations.

The problem is that so many of the world's forests, and so much new planting, are not being managed with these methods, despite centuries of forest devastation and restoration. Much more public pressure is needed to push the adoption of forest-saving and regenerating practices—and not only in the devastated rain forests of the Amazon and Indonesia. The ravaging of the Amazon in particular has received intense media coverage, but destruction of the rain forests of Canada is also rampant. Rain forests in North America? Yes, Canada's British Columbia is home to the largest remaining temperate rain forests on Earth, and large swaths are being clear-cut. Meanwhile, forests in the southeastern U.S. are also being decimated, with much of the wood being turned into pellets for wood-burning heating systems in Europe; according to environmental scientist William Moomaw, the loss of forest canopy there is greater than any place on the planet.

The good news is that circular economy initiatives are showing the way, informed by discoveries about the extraordinary complexity of forest ecosystems. Forester Peter Wohlleben, who manages a beech forest in Germany's Eifel Mountains, beautifully showcased the wonders of forest ecology in his book *The Hidden Life of Trees*. "Trees unite to create a fully functioning forest," he writes, whose "whole is greater than its parts." Researchers have found that beech trees share with one another the sugar they create by photosynthesis, through an intricate network of roots and the mycelium woven throughout the soil, equalizing nutrients among them. Higher producers are providing for lower ones in a form of tree socialism. One might expect that trees would only help other trees of their own species, but not so. Forest ecologist Suzanne Simard discovered that the towering Douglas firs in the old-growth forest of British Columbia, which hold on to their dense green needles year round, send nutrients to nearby deciduous birch trees after they've dropped their leaves in fall. Birch trees return the favor by sending over to firs growing in shady areas some of their own sugar produce.

Adaptive wizardry is probably exhibited no more gloriously on the planet than by rain forests. They are said to account for about 50 percent of global biodiversity, home to such richness of species that it's been impossible to reach a total count, but estimates go as high as 50 million. Just one hectare of rain forest might support many hundreds of tree species, whereas in old-growth deciduous forests, the count might be twenty or fewer.

Forests have been called the planet's lungs: after sucking in carbon dioxide, they breathe out oxygen and send ample quantities of carbon underground as rain carries it from decaying fallen limbs and trunks down into the soil. One estimate of the potential of forest carbon sequestration is that current forests worldwide are capable of absorbing about a quarter of all human carbon emissions—if so many weren't being logged and burned. Finding ways to prevent the raging fires that have devastated so much forest in recent years, particularly in the western U.S. and Australia, must be a priority in the climate change battle. Thankfully, forest scientists have produced powerful insights into why massive forest fires are becoming more frequent and ferocious, and also into the best ways to combat not only fires but also the rapacious logging and intentional burning to clear land for farming that are additional major contributors to depletion.

Some forestry experts argue that leaving our older forests entirely alone is optimal, while others advocate their active management, including selective tree harvesting. Some studies show that younger forests are the better carbon sponges, so planting as many more trees as soon as possible should be the focus. To this aim, in

2020 the World Economic Forum launched 1t.org, an organization to support the planting of 1 trillion trees, aiming to accelerate the work of organizations such as the Trillion Trees Initiative, which runs the Plant-for-the-Planet app, which lists tree planting organizations all over the world that take donations. Bold as that target is, it leaves open a challenging question: How can we save and grow forests optimally? Some experts caution that the land for such new growth is limited, and simply planting for maximal carbon sinkage is misguided. Cultivating tree plantations for harvesting is a prime means of taking the burden off natural forests, but they too present challenges, sometimes falling prey to devastating blight.

Mark Ashton of the Forest School at the Yale School of the Environment is one of the world's leading experts on forest growth and cultivation. Ashton was born in the Sultanate of Brunei, on the Southeast Asian island of Borneo. His father was forest botany legend Peter Ashton, the first to catalog all the tree species of the island's once lush rain forest—then so forbidding that Ashton repeatedly almost lost his life on expeditions, but now almost entirely destroyed. The natural profusion of tree species—Ashton counted some three thousand—has been replaced by monotonous palm oil plantations.

Peter was also a pioneer of the comparative study of biodiverse forests around the world to learn about best practices for preservation and renewal, which Mark has specialized in. Having studied plots for over thirty years in Sri Lanka, India, and New England, Mark advocates for a portfolio management approach to optimizing our preservation and cultivation of forests—advice aligning beautifully with the circular economy philosophy of pursuing myriad methods that will best serve both the needs of people and the planet. His advice also aligns with the fundamental premise of circularity: that following nature's lead is the surest course.

Much of the cultivation of new trees for carbon sequestration thus far, Mark points out, has involved planting only a few, or even just one species; it's thought that because they grow particularly fast and are long-lived, this method will suck up the most carbon the quickest. But Ashton's research has shown the approach produces forests highly vulnerable to blight. With so many trees of one species tightly packed, some pest can come along and hop so easily from one to another target tree that "they destroy the whole clan," as Ashton puts it.

His study of how natural forests become resilient shows that a diversity of tree species should mature together—which for me calls to mind the remarkable diversity of trees that have sprouted in the spontaneous natural woods of Freshkills Park. "You have to delegate the decision making to nature," Ashton says, to make both forests and tree plantations "capable of withstanding all potential assaults" by insects, tree diseases, and droughts, as well as selective human harvesting. In his words, "You're maximizing resilience, not necessarily carbon storage"—ultimately the surer route to optimal carbon sequestration.

What of the ferocious fires in California and Australia, I asked. What's been fueling them?

Another well-meaning but misguided method: a policy of combating all forest fire. Protecting trees from fire so zealously has led to "higher amounts of carbon stored in these forests than should ordinarily be there," Ashton explained. Excess carbon has acted as fire propellent, in concert with drought and increased populations of invasive insects due to climate change, leaving many more trunks and branches on the forest floors for kindling. We should be practicing instead, Ashton says, the fire-prevention method employed for eons by Native Americans and indigenous Australians—controlled, low-intensity, and frequent burns.

In territory in northern Australia, members of the aboriginal population, by agreement with the government, have done just that to great effect. Since the program's inception in 2013, the massive fires that had been breaking out in this region prior to the implementation of the program have been reduced by half, and greenhouse-gas emissions cut by an estimated 40 percent. Participants in the program have been incentivized with cash rewards through a cap-and-trade scheme, with payments so far totaling $80 million—an important economic boost for a region in which unemployment is high.

Ashton stresses that such economic incentives are crucial to forest preservation, not only in the Amazon but in so many forests where clearing for lumber and agriculture continue. Devastation of forests won't likely be stopped until the economic incentives for preservation are greater than those for destruction. That's true not only in the developing world but also for the plundering of forests in the U.S. and Canada. If even in such wealthy nations, and despite considerable objections from environmentalists, economic payoffs have perpetuated forest destruction, then expecting better preservation in economically challenged regions without compelling monetary compensation is simply wishful thinking.

One of the most promising approaches to providing that compensation is by accounting for the value of forest ecosystem services, which has come a long way since Thomas Graedel and his cohorts spearheaded the practice. In addition to being the planet's lungs, forests are her water-filtration systems, playing a leading role in providing us humans, as well as wildlife, with potable water. A single mature tree in a temperate forest might take in, purify, and release a hundred gallons of water a day, and a rain forest tree perhaps twice that. So prodigious is the water filtration of the Amazon that it's said to account for 15 percent of the planet's drinking water. Forests also help prevent flooding and erosion; one study found that trees planted along streams that feed into the ocean increase the yields of fish in fisheries and oysters in oyster farms, because their fallen leaves raise the acid level of the water and boost the growth of the plankton that sea animals feed on. Paying communities for forest stewardship has shown impressive results and is catching on globally, with a recent assessment of such programs finding that about $36 billion was being paid annually. An analysis of one program in Mexico showed it reduced the destruction of trees by about 38 percent. In California, a "tropical forest standard" has been formalized, which stipulates best practices for management of the Amazon rain forest and allows local and national governments who prove their adherence to them to sell carbon credits to California companies.

Preserving old-growth forests and planting new trees must be combined with forest culling, tree farming, and payment schemes—and we can take heart that if they are, forests globally will thrive. Damage done can be reversed. Consider that after colonization of the U.S., over half of the forestland of New England was cleared for farming; now, with most of those farms having been abandoned, 80 percent of the region is again covered in forest. We might think of the Amazon as primordial, populated only by small tribes, but in fact it was inhabited for thousands of years by large societies who cleared much of the land for farming and eviscerated many tree species to make way for the breeding and harvesting of nut, palm, rubber, and cocoa trees. The Amazon today is lush in a much different way from before these interventions. Carolina Levis, one of the researchers who made this discovery, highlights that it shows that "human influence can enrich the Amazon." That is what many of those in indigenous communities living in the Amazon are doing, making important contributions to its preservation.

The Ceibo Alliance is a nonprofit founded by members of the indigenous Kofan, Siona, Secoya, and Waorani peoples who live in the Ecuadorian Amazon. The group works against the clearing of the forest for cattle raising and palm oil and rubber extraction. Co-founder Nemonte Nenquimo, named one of *Time*'s 100 Most Influential People of 2020, wrote powerfully in a piece in *The Guardian* titled "This Is My Message to the Western World—Your Civilisation Is Killing Life on Earth": "In all these years of taking, taking, taking from our lands, you have not had the courage, or the curiosity, or the respect to get to know us. To understand how we see, and think, and feel, and what we know about life on this Earth." The work of the Ceibo Alliance is, at last, helping to correct that gross indignity. Nenquimo led the fight in a winning lawsuit against the

Ecuadorian government for violating her people's rights to prevent oil exploration in their territory of the Amazon. In partnership with Amazon Frontlines, an international group of environmental activists, lawyers, forestry scientists, and anthropologists, she and the alliance are crafting additional legal strategies to defend the rights of the indigenous forest peoples as stewards of the land. Many indigenous forest dwellers have also collaborated with the Rainforest Alliance, another international organization, to find ways to sustainably farm in rain forests, according to ancient practices that preserve biodiversity, as well as to work in other forest preservation jobs, and the alliance reports that since 2011, these indigenous partners have collectively earned $191 million through this work. They are helping prove that forest preservation is economically as well as environmentally superior.

As Suzanne Simard says, "As complex systems, forests have an enormous capacity to heal." We in the public all around the planet can do a great deal to assist in such efforts to protect and heal forests. We can buy only paper products made from sustainably managed forests, labeled with the FSC certification. We can purchase reusable folding cups, which are readily available online; we can switch to reading the news only online. As plastic bag bans go into effect, hopefully in more and more areas, and stores replace them with paper bags, we can bring reusable bags. We can look for all sorts of ways to cut down on our paper use both at work and home. My book editor told me that given the reams of paper she has to print out in order to edit manuscripts, she prints on both sides of the paper. When it comes to paper and cardboard packaging, we can buy brands that use recycled content and bring refillable containers to stores that sell some items in bulk. Of course, being rigorous about sending all of our used paper for recycling is key.

We can donate to any number of forest relief and regrowth projects, such as Stand For Trees, the Congo Basin Rainforest Project, and the Amazon Forest Protection Project. We can also petition our government officials to press for better enforcement of forest regulations and to pass stricter ones, and donate to organizations such as Conservation by Coalition, which help improve forestry regulation in the U.S.

As we can see from the irrepressible phoenix regeneration that's turning once despoiled acres of the Fresh Kills Landfill into a rich, forest ecosystem, if we give forests a chance, they will astonish us with their resilience and support.

5

Greener Grocery

SOON AFTER I JOINED the Bloomberg administration, I learned that the city had a "ratologist" on staff named Robert Corrigan. Right away I made an appointment to speak with him. I had something important to discuss, but I also wanted to meet a real-live ratologist since I hadn't known such a profession existed. I wanted to make one of my first initiatives a food-waste collection program, and I thought he could be a big help in gaining support.

Corrigan holds a PhD in rodentology. He had planned to become an oceanographer, but just one lecture by a visiting pest management professor had him hooked. As a graduate student, he lived for thirty days in a barn brimming with rats so he could observe every little detail of their nocturnal festivities. Since then, he's been a great admirer of "a spectacularly beautiful mammal by all measures," and sought to control rat population growth rather than kill them all off. He keeps up a lively Twitter feed raising rat awareness with observations like, "You know, you never heard James Cagney say you dirty mouse." Corrigan is now in his midsixties, and with his wire-framed glasses and all-business manner might more readily be taken for an accountant than the eager leader of what he calls "real-deal nighttime rat safaris."

Teaching clients his "Sherlock Holmes approach" to discovering rodent thoroughfares and hiding places, his safaris disclose how unfortunately hospitable American cities are for them. He points to tiny holes between bricks in building walls, which he dubs rodent condominiums, layered with torn bits of paper and larded with food scraps. City streets become veritable rivers of rat locomotion after the city's restaurants, grocers, and apartment buildings toss mountains of food-filled garbage bags onto curbs every evening.

Because most cities have residents include their food waste in the trash they set out at the curb, food foraging is a breeze for rodents. Corrigan points out that many of New York's trash cans have facilitated their feasting. For years most have been open-topped and wire-meshed, making them optimal for climbing in and out of; he calls them rat ladders. Cities don't have a rat-control problem, he says; they have a food-waste problem.

Environmental policy often gets caught up in heated political bickering. However, one policy I knew would align all groups: reduce the city's rat population. Of course, separating our food waste from our trash so it can be composted and turned into soil nutrients, or used to make clean energy through anaerobic digestion, should be appealing on its own merits—not to mention that it reduces landfill disposal fees. But I knew that demonstrating that the collection program would cut the number of rats scurrying around sidewalks and subway tracks would inspire unanimous approval. Robert Corrigan proved invaluable because he endorsed special residential bins for food waste, emphasizing that they would eliminate the rats' ability to smell any food, thereby also eliminating their interest in our neighborhoods. We got speedy approval for the program and local community support.

While making good use of food waste is one vital component of developing a circular food system, we also need to dramatically reduce the volume of food waste generated. The amount of perfectly good food thrown out every day around the country, and much of the rest of the world, is an inexcusable travesty. In the U.S., an estimated 40 percent of food produced is wasted, and the global estimate is 30 to 40 percent. *Consumer Reports* estimates that the average American throws out a pound of food, or 1,250 calories, a day. American households are said to account for 43 percent of overall waste, restaurants 18 percent, groceries 13 percent, institutional food services such as hospitals and schools 8 percent, and farms 16 percent. Consider that the average American throws out ten times more food than average consumers in Southeast Asia and sub-Saharan Africa. In fact, food waste constitutes the largest portion of the waste stream sent to landfills, at just over 20 percent. Paul Hawken's Project Drawdown ranked reducing food waste as the third most effective means of reducing greenhouse gases in the atmosphere.

So the first question to ask about how we can make the food economy circular is why on earth on are we trashing so much good food? One reason is that we've been convinced to do so.

Buyers Be Less Wary

"Sell by," "Use by," "Best by," "Best before," "Best if used by," "Freshest before," "Expires on." Much of the food we purchase is stamped with one of these labels; often with both a sell-by and use-by date. The dates look official and precise, and are presumed to be based in science. They are not. What they are is a leading cause of the disposal of approximately $29 billion of food every year in the U.S. alone—food that tastes just fine and is safe to eat. Who comes up with them? Many are calculated and stamped on products by manufacturers, while some are affixed by retailers according to supplier suggestions. How do they come up with them? For many foods, manufacturers leave samples to rot in labs and then factor in an expected amount of travel time to the grocery shelf to guesstimate dates. The temperatures foods will be exposed to on their journey through the supply chain, from the grocery store to the home and for what period of time and other factors involved in spoiling, such as humidity, are far too variable to allow for precision. The dates aren't even estimates of when food will actually go bad; instead, they're estimates of when food will look, smell, and taste optimally good. They're not required by federal law, as many consumers think (with the exception of baby food, for which a use-by date is mandatory). Though forty-one states have regulations requiring dates on some products, such as milk, they're a muddled hodgepodge state to state. There is no nefarious intent by producers. The cause is often inefficient and antiquated supply chains that don't provide producers and retailers the data necessary to properly and consistently advise their customers. The result is that they do their best to put on a date that approximates the freshness of a product. The current accuracy of that approximation is a loss of billions of dollars of food coupled with billions spent to send perfectly edible food to landfills.

The enormity of the food-waste problem in the U.S. does not end with food wasted due to inaccurate date labeling. On top of it is good food tossed by grocers, restaurants, and consumers judging it based on its appearance as opposed to actual taste and freshness. For some grocers, some of their waste is by intention. They throw out considerable amounts of fresh food every night because they plan to stock more than they're going to sell. The general belief is that shoppers like to see large mounds of avocados and tomatoes, peaches and pears, all beautifully shaped and bruise-free, and that we buy more when that's what's before us. Overstocking is so common with displays of fish that a study found that 26 percent of fish stocked in the U.S. isn't sold, despite the fact that 90 percent of fish sold in the country is imported, much of it traveling many thousands of miles from Asia.

Doug Rauch, former president of Trader Joe's, explained the thinking by grocers this way: "If a store has low waste numbers, it can be a sign that they aren't fully in stock and that the customer experience is suffering." Stores suffer considerable loss in profit due to that belief, which, with margins for the typical grocery chain coming in on average between just 1 and 3 percent, they can barely afford. Yet, when Stop and Shop conducted an analysis of the effects of doing away with "pile 'em high, watch 'em fly" stocking of perishables, it found that customers actually preferred the smaller displays because the items were much fresher; in fact three days fresher, generally. The chain has saved $100 million annually from the change. Grocery stores are learning that this myth that they religiously followed for years was completely unsupported by any data or reputable studies. The only data is on their historic P&L statements showing the millions of dollars they lost buying more than they knew they could sell and then paying to dispose of it in landfills.

Much of food waste comes down to the way we think about fresh or still packaged food; how we value it, and what we value about it. Perhaps no one has a better understanding of this, or an appreciation of what a hearty and healthy bounty of food we trash, than the modern-day foragers who call themselves freegans. One well-organized group in New York City, which runs the website Freegan.info, is dedicated to raising awareness of the food-waste travesty and holds monthly freegan tours. A reporter friend of mine joined their tour and called me afterward for some insight about what she'd witnessed.

The tour met at 9:00 p.m. on Manhattan's Upper West Side, about twenty people across a wide range of ages. A number of the group were first-timers like my friend, but many were dedicated foragers who went out several nights a week. Several members showed up with laundry carts, which surprised her. Were they really going to find *that* much fresh and still packaged food?

The tours start at nine because most grocers put their trash bags out by then, but the garbage trucks don't come by for pickup for another couple of hours. One of the members explained the strict rules of their foraging: Bags must not be torn open, but instead unknotted, as the group is intent on not causing their own trash problem. Anyone who makes a discovery should call out what was found so others with interest in those goodies can share them. And in a lovely gesture, the leader asked who needed gloves, offering up a bag full of mismatched mittens she'd collected. My friend grabbled a set, given that they were going to be riffling through bags of garbage.

"Ron, the bags were triple-knotted," she told me. "Why is that? They were so hard to open." Stores want to make it hard for foragers, because they don't like customers seeing them going through their bags.

Some of their discoveries were surprising: she found a whole bag full of romaine leaves, which close inspection revealed were perfectly crisp and still wet from washing. "Why on earth would they throw them away?" she asked. Customers prefer smaller heads of lettuce, so none will go bad in their fridge; so the stores pull off the outer leaves and discard them.

They had also found a bag full of boxes of grains in front of a natural foods store, which included a pricey purple barley, a nearly extinct heirloom variety. Dried grains last a near eternity; why would they be thrown out? Sell-by dates aren't the only arbiter of shelf-culling. Inventory clearing is also common. In fact, that's often required by brands who have elaborate deals with stores dictating what they stock, where, and for how long. What was particularly disturbing to her was that the boxes had all been slashed open. I explained that stores routinely have staff slash packages that way, as additional forager deterrence. Her account included a veritable grocery cart full of perfectly fine fruits and vegetables and a massive bag of fresh bagels. Those are staples of store waste, along with bread, yogurt, cheese, and plenty of meat and fish.

The next night, a group of fifteen got together to enjoy their findings. As they peeled and chopped and stirred, they blended several rounds of fruit smoothies, a freegan staple. Their haul yielded two rice dishes and one with the purple barley, loaded with tomatoes, carrots, turnip, tofu sausage, bean sprouts, and broccoli—with a touch of heat courtesy of a mystery Korean sauce, with no English labeling, from an Asian grocery where they had found dozens of bottles—served with a large salad, a big plate of bread, and, somewhat oddly, a can of jellied cranberry sauce.

From Healthy Bounty to Food Insecurity

Anthropologists who've studied the few remaining hunter-gatherer societies and archaeologists who've uncovered fascinating remnants of their ancient past have provided a compelling portrait of the foods they foraged. They are said to have made quick work of it, gathering and hunting an estimated seventeen hours a week per group member, so a little more than two and half hours a day. Their diet, for the most part, comprised what Michael Pollan recommends: "Food, not too much, mostly plants"—"real food," that is, as opposed to the "edible food-like substances" that food scientists have concocted.

Analyses of ancient hunter-gatherer bones suggest they were largely free of most of our present-day maladies, from obesity to diabetes and heart disease. Descriptions of their health are so appealing that they beg the question: Why would people have decided to settle down and become farmers? The work of cultivation was much harder, and the early agricultural crops were less nutritious, as indicated by the skeletal remains of early farmers, who were shorter and had less healthy bones and teeth. One idea is that the transition to cultivation was driven by beer. The grains for making beer were the first to be domesticated—wheat, barley, rice—but most food archaeologists thought bread was the first grain-based human-created food, until botanist Jonathan Sauer suggested that the natural fermentation of barley into alcohol spurred grain cultivation. A debate raged, but as recounted by scholars Solomon Katz and Fritz Maytag, the decipherment of an ancient clay tablet from Sumeria, a culture so besotted with beer it worshipped a goddess of brewing, may provide the answer. Detailing a recipe for a favored brew, the tablet explains that bread was made by the Sumerians as a way of storing the raw ingredients for beer. So, perhaps the beer-or-bread question is a chicken-or-egg problem. At any rate, whichever came first, it's not hard to imagine that brewing beer was much welcomed compensation for farming's hardships.

The greatest travesty of our copious food waste is that even as so much perfectly good food is thrown away, so many people are suffering from food insecurity. Recall that an estimated 12 percent of American households have been found to be food insecure, and that was before the COVID-19 pandemic led to increased stresses on food supplies. *Forbes* reported in May 2020 that the number had at least doubled due to the outbreak, with estimates of the rise ranging from 22 to 38 percent. That's despite a host of organizations devoted to collecting food that would otherwise be thrown away from restaurants, food-service providers, and grocers, and redistributing it to the needy. This vital work was deservedly spotlighted for the first time as food shortages hit due to the coronavirus. Yet it's estimated that only 5 percent of the food that could be distributed to those in need is distributed.

A key problem is that participation by food purveyors is far too low. For example, a mere 2 percent of food that could be donated by national restaurant chains in the U.S. is donated. One explanation is that potential donors fear legal liability and brand damage if food they've offered turns out to be bad and causes illness. Yet in the U.S., the Bill Emerson Good Samaritan Food Donation Act, passed in 1996, protects donors from any liability as long as donations were perceived to be "wholesome food or an apparently fit grocery product." The underlying issue is, again, one of consciousness. Throwing food away is seen as easier than making arrangements for its dispersal, but the hassle-free option comes at a considerable cost for disposal alone. An estimated $1.3 billion is spent by the food industry in the U.S. to trash haulers, even while many food banks are forced to purchase the food they serve. Even school systems, which have become lynchpins of food provision for millions of children who would otherwise go hungry and thus are intimately aware of the food insecurity problem, dispose of vast quantities of food that could be donated. The USDA's National School Lunch Program is said to waste the equivalent of $5 million spent on food each day, for a total of $1.2 billion a year.

Finding Circular Solutions

Fortunately, a legion of innovators are devising brilliant solutions for closing the loop on food waste, some of which are blazingly simple and low-tech, and others that are highly tech savvy. One solution that requires nothing but initiative was implemented in schools in Oakland, California, by the district's sustainability manager, Nancy Deming. She has required all schools to put receptacles for composting, recyclables, and unopened items, such as cartons of milk and fruit cups, next to trash cans in their cafeterias. She's also instituted share tables, on which children are asked to deposit any food that comes with their meal that they don't want to eat. Uneaten food is sent to homeless shelters and food banks. So simple, yet so effective.

As for high-tech solutions, a host of apps have been developed to facilitate redistribution of food from restaurants, food-service companies and caterers, groceries, and homes. One of these, Food Cowboy, allows those with food they want to donate to send an alert describing what's on offer, which goes out to the food banks and others who've registered with the app, and donor and recipient then make arrangements for delivery. Another app, Food Rescue US, uses sophisticated algorithms to analyze optimal locations for the delivery of food that donors send word about, then arranges for volunteer drivers to pick it up and get it there.

Another high-tech solution is food sensors that provide real-time information about how fresh foods are, correcting for the imprecision of sell-by and use-by guesstimates. Israeli company Evigence invented a sensor easily applied to food packaging, a green circle whose outer rim begins to change to red, section by section, according to the number of hours the food will still be good—finely calibrated to detect freshness, based on elaborate calculations made of temperature readings over time as food spoils. Researchers at Imperial College London created paper-based electrical gas sensors that detect a range of gases emitted as food begins to spoil. They're embedded with a microchip so that consumers will be able to hold their smartphone up to a sensor and get an exquisitely accurate reading. At a cost of only two cents each, they may well be embraced at scale.

While better monitoring of freshness on store shelves and in our fridges will be a marvelous advance, I'm especially excited about an innovation that promises to actually keep perishables fresh longer: a microscopic food coating that is edible.

As soon as I met Adam Behrens, the CEO of Mori, I knew he was the kind of guy whose parents never had to worry about his future. When we met, he had recently made the bold decision to leave a prestigious post at MIT's Langer Lab, where as a postdoc in biomedical science he'd been conducting research to develop vaccines and to add nutritional fortification to food, aimed at improving health in the developing world. His graduate research involved developing materials to spray on skin to stop bleeding. He'd even won funding from Bill Gates.

Adam and his team had launched Mori, a company that provides a microscopic and edible coating that stops food from spoiling, with lettuce growers the prime contenders for early adoption. The invisible, edible, and tasteless coating is made from silk molecules, extracted from silkworm cocoons, using just a mixture of salt and water. The recipe was developed by Professor Fiorenzo Omenetto, who runs the silk lab at Tufts University, in partnership with Professor Benedetto Marelli, a materials scientist at MIT. They'd been working on creating coatings to protect medicines and vaccines when Marelli got the idea that the silk coating might also work to preserve foods. After spraying a microscopic layer, the thickness of just two blood cells, onto a strawberry, he returned several days later to find the strawberry was still fully fresh. Testing with a number of foods has shown that the coating can extend freshness by two to three times, and that's without refrigeration. As a postdoc in the lab, Adam thought the potential was so exciting that he threw himself into building Mori, despite having no prior entrepreneurial experience.

He also had no prior experience with agriculture, and when I asked him what the biggest eye-opener was for him about the farmers he's met with and their operations, he responded immediately: "How sophisticated their growing methods are; they really have to be scientists." Many have installed sensors in their fields to measure soil nutrition levels, refine fertilizer and pesticide application, and gauge moisture to better control irrigation. They're constantly scouring readings from their fields on their laptops, and poring through trade magazines and websites featuring findings from agricultural research. Most are regularly experimenting with new seeds and growing advice from agricultural scientists. But despite the most rigorous efforts, any given farmer in any given growing season might face calamity. Adam recalled somberly strolling through a cherry orchard that had been utterly devastated by hail; and for lettuce growers, E. coli is a particular terror.

The plan for Mori is to work initially with farmers, to supply the coating for them to apply to crops right after harvest, and then perhaps to expand sales to packagers and groceries for coating meats and fish as well. For lettuce, this would help prevent substantial waste in grocery stores and in the massive volume of packaged bags of salad greens that never make it to retailers. Day after day at the Salinas dump, bags stamped with sell-by dates still two weeks away are unloaded by the local packagers. That's not enough time to assure that they'll reach stores with adequate shelf life to meet grocers' standards, given that they're traveling so far and will typically spend time in distribution centers as well before making their ultimate journey to stores. Though no hard data is available about how much lettuce, as well as other produce, is wasted even before then, disposed of straight out of the field due to overplanting, the National Resources Defense Council estimates that many growers intentionally overplant by about 10 percent. Adam hopes that Mori's ability to keep food fresh longer will convince farmers to grow less. Mori will obviate the need for some refrigerated transport, which would help bring down carbon and methane emissions by reducing the energy used to get the produce to stores. Additionally, using Mori will enable farmers and grocers to reduce the ridiculous amount of plastic wrap currently used to maintain freshness and will extend the amount of time food can remain in the fridge and on the shelf in people's homes.

Mori is one of a number of food coating start-ups, each promoting its own unique product, and they're starting to gain great market momentum. Apeel Sciences, which produces a plant-based coating, has run trials for grocery giant Kroger with avocados, reportedly preventing rotting for as long as a month. The company states that coated apples and blueberries also stay good for a month, while limes hold up for seven weeks.

Back to Farming's Future

The industrial agricultural approach to growing that's dominated farming for most of a century has caused so much soil degradation and has taken the use of fertilizers and pesticides to such extremes that it's reaching the limits of productivity even as the world's population is estimated to increase by 2 billion by 2050. The result is that the planet is barreling into a future of possible wide-scale famine, even as farming contributes to producing the conditions for famine. It didn't have to be this way; the organic movement developed hand in hand with industrial agriculture, and it provided powerful proof of the superiority of natural growing as early as the 1920s.

In 1905, the thirty-two-year-old English botanist Albert Howard and his twenty-nine-year-old wife, Gabrielle, a plant physiologist, embarked on a great adventure. Shortly after their wedding, they traveled to the agricultural hub of Pusa, in northeastern India, to conduct research at the country's premier agricultural institute on growing heartier wheat and cotton. Both with Cambridge degrees, they might have been expected to view the peasant farmers of the area as backward. Instead, heading out into the farmers' fields, they recognized how brilliantly the farmers followed ancient techniques of what we now call organic growing. A friend of Albert's who worked with the couple at the time recalled that he "was always wont to say that he learned more from the farmer in his field than he did from text books."

Chemical fertilizers had been championed since the mid-nineteenth century as "wonder drugs" for cultivation, especially nitrogen, potassium, and phosphate. The surface soil of the farmers' fields was quite low in these nutrients, and yet they produced an impressive bounty. How? They planted a rotation of deep-rooted crops in fields and surrounded the fields with fruit and other types of trees, and the root systems of all of those plants became natural fertilizers, pumping nutrients into the soil and distributing them widely. Studying their soil also revealed that plentiful microbes feasting on the minerals in the soil were additional sources of natural nutrients, which chemical dousing killed off. The farmers' soil was alive, generating its own organic fertilizers. Preserving the life of soil, and even better, enriching its content of organic matter, the Howards realized, was the key to both high production and healthy production. Chemical fertilizers were, by contrast, killing soil. With that realization, the Howards worked to help the farmers further enrich their soil by developing ancient Chinese and Indian techniques of creating humus, most often now called compost, into a full-fledged science of composting.

The Howards' discoveries did not go unnoticed by the global agricultural community. In 1931, when Albert journeyed by ship around Africa on his way back to England after Gabrielle had sadly passed away, he was invited to visit coffee growers in Kenya who had embraced the methods. So had tea growers in Assam and Ceylon. In England a few years later, Albert was joined by many of the country's most prominent agricultural innovators, including earls, viscounts, and lords, at a high-profile symposium about organic methods. One of those noble gentry, Lord Northbourne, aka Walter James, coined the term "organic farming" in his book *Look to the Land*, which was published in 1940. Albert Howard wrote a flurry of books in the 1930s and '40s promoting organic growing methods, most notably *An Agricultural Testament* and *The Soil and Health*, which were widely read and acclaimed. As his second wife wrote in a tribute to him after his death in 1947, they also "aroused first contempt and then a frantic opposition" from the chemical purveyors and enthusiasts.

In one of those ironic coincidences of history, the year of Albert Howard's death was also the year that, as Michael Pollan highlighted in his book *The Omnivore's Dilemma*, a massive plant built by the government in Muscle Shoals, Alabama, during World War I to crank out nitrate explosives, was converted to cranking out chemical fertilizers instead. The Muscle Shoals plant and nine other such facilities that the government ran were sold to private fertilizer producers who massively increased fertilizer production, driving prices down. At the same time, government policies incentivized farmers to purchase them in much higher quantities so they could adopt a new model of cultivation that was the antithesis of the organic system Howard spearheaded.

In 1944, Norman Borlaug, a thirty-year-old specialist in crop breeding, traveled to Mexico to help its wheat farmers improve their results. He'd grown up on a small farm in Iowa and done his agricultural studies during the horrifying Dust Bowl years, and his belief that a lack of technological prowess had caused the catastrophe drove him to experiment with creating heartier hybrid plants and boosting their growth with refined mixtures of synthetic fertilizers. While in Mexico, he created a dwarf wheat hybrid that grew much shorter than most wheat varieties, making the stalks more resilient, and produced a much higher overall yield. Word of the achievement ripped around the agricultural community.

Populations were booming all around the globe in the years after the war. In the U.S., the population soared from 130 million in 1940 to 151 million in 1950. There was widespread fear that waves of famine loomed unless yields of the major food crops—in addition to wheat, corn, soybeans, and rice—could be greatly increased. The USDA and major world organizations promoted Borlaug's hybrid seeds and synthetics approach as the solution, and it ran roughshod over the emerging organic methods.

A horrible boomerang effect of the adoption of the Borlaug approach in the U.S. was that the spike in yields for corn, wheat, and soy led many farmers to devote their land entirely to growing only one or two of these crops. More land was also given over to lower-quality grains to feed livestock, called "commodity crops," with farmers and ranchers bringing their animals in from fields, where they'd grazed for food, into densely packed feed lots. How densely packed? Most state guidelines allow for 10,000 hogs, 125,000 chickens, and 1,000 cows on a ten-acre parcel. For comparison, consider that the recommended number of acres per cow raised by grazing is 1.8, which translates to about 6 cows for 10 acres. What's more, because cows evolved to eat grasses, not corn, they must be fed antibiotics to keep them healthy.

Today, only 27 percent of farmland in the U.S. is dedicated to producing food for humans, with the rest growing the commodity crops. The big three grains, along with cotton, have so dominated farmland in the U.S. that only 2 percent of cultivated land grows vegetables and fruits. Meanwhile the varieties of produce available have been decimated. At the beginning of the twentieth century, American farmers were growing 544 types of cabbage, 497 types of lettuce, 408 varieties of peas, and 408 tomato options. As *Fast Company* summarized the shift, "in 80 years we lost 93% of variety in our food seeds."

After the war, many more farms also mechanized, with much more efficient harvesting equipment. The combination of higher yields, concentration on a few crops, and mechanization led to huge excesses of the major crops flooding the market, as Rachel Carson had noted. Prices plummeted and, for many farms, bankruptcy loomed. So the government introduced so-called price support policies, which included paying farmers to leave some of their land fallow and subsidizing crop insurance, which compensated farmers for losses, creating the perverse allure of "farming the subsidy." To prop up commodity prices, the government bought vast quantities of grains that it stored away in enormous silos. Dairy farmers also received support, with the government buying vast quantities of milk, butter, and cheese. In a bizarre testament to how absurd the system became, those dairy products were stored in freezers and cooling rooms installed in an old limestone mine and several underground caverns on the outskirts of Kansas City, Missouri.

The absurdity of the system has been zealously defended by Big Ag corporate behemoths. Relentless consolidation has left just six giant firms dominating the sale of "inputs," meaning seeds, fertilizers, and pesticides: Dow, Monsanto (which recently merged with and took the name of Bayer), DuPont, BASF, and Syngenta. Just three firms now dominate the purchases of the big four crops from farmers: Cargill, Archer Daniels Midland, and CHS Inc. The interests of these giants are in selling farmers more and more inputs, at increasing prices. One result is that just 1 percent of farmland in the U.S. is devoted to organic growing, even as demand for organics is so high that we import billions of organic products from overseas every year.

Meanwhile, most farmers are increasingly squeezed, with razor-thin profits even in good years and all too often suffering losses, with subsidized crop insurance often inadequate to make up for them. Coverage of the industrialization of farming has led to a widespread perception that most farms are enormous corporate-run operations, when in fact 97 percent of farms in the U.S. are still family owned, and most are smaller than a thousand acres. However, out of total annual government subsidies of between $15 and $20 billion annually over the past decades, the vast majority of funds have gone to the small number of huge corporate farms. From 1995 to 2017, for example, the largest 10 percent of farms received 77 percent of a total $205 billion. For family farms, meanwhile, bankruptcies have loomed, with so many facing desperate straits that the suicide rate among farmers in the U.S. has spiked to 50 percent higher than that of the general population.

How could such an unjust system prevail? It is due largely to the diminishing political clout of farmers and a parallel increase in the lobbying power of Big Ag. The number of farmers in the U.S. began plummeting in the 1950s, from 30.5 million before World War II, which was 23 percent of the population and the single largest voting bloc, to 2 million today. Contributions to politicians' campaigns more than compensate for any blowback due to farmer discontent.

One of the pernicious revenge effects of this sad affair is that with the contraction in the number of growers of fruits and vegetables, their prices have risen. This was a major factor in food-processing companies like Kraft and General Mills replacing these natural sweeteners with sugars and fats derived from inexpensive grains (most notoriously corn syrup). The results for diets and health have been tragic.

As Dr. Mark Hyman reveals in his book *Food Fix*, the larding of processed foods with sugar, salt, and fat has led to the epidemic of obesity not only in the U.S. but around the world, as well as to large increases in diabetes, heart disease, and other ills. A shocking recent study found that participants given processed foods ate on average 50 percent more calories a day than those given a healthy, unprocessed diet. What's more, those eating the processed foods ate much faster, because signals that travel from the gastrointestinal tract to the brain to indicate fullness were circumvented. That is by no means accidental. Processed food companies have gone to great lengths to get us to eat more, faster.

Food scientists working for Kraft and other industry leaders have discovered many secrets to making foods addictive. The science of food flavoring has become incredibly sophisticated, making use of high-tech equipment including spectrometers, gas chromatographs, and headspace vapor analyzers (whatever they are) to gauge how effectively the flavors they inject into foods are hooking eaters. Injecting flavor is necessary because otherwise the foods they've so elaborately processed would be almost entirely devoid of it. As Michael Moss relates in his often shocking book *Salt Sugar Fat*, by 1960 this army of "flavorists" had created fifteen hundred artificial flavorings. Yet, even so, the three flavor boosters in his title are by far most responsible for the insatiable hunger for processed foods today. In a revelation that staggers the mind, he shares that they learned how to make processed sugar two hundred times sweeter tasting than its natural counterpart. They also learned how to determine the "bliss point," which Moss explains is "the precise amount of sweetness . . . that makes food and drink most enjoyable." So enjoyable because, as Moss reports, food scientists also learned that the brain reacts to sugar much as it does to cocaine. Perhaps the most pernicious finding Moss describes, though, is about "vanishing caloric density," which is the phenomenon that causes food that quickly melts in our mouths to register in our brains as having no calories. As food scientist Steven Witherly told Moss, "You can just keep eating it forever." Which explains the voracious quantities of potato chips and puffs like Cheetos we can so quickly scarf up.

Mark Hyman writes that to get out of this bind, food cultivation and distribution innovations must "produce real food at scale" through a "reimagined food system from field to fork and beyond." That is precisely what circular food economy innovators are well on the way to achieving, beginning by looking backward to move forward.

Farming That Regenerates

I reached Gabe Brown on his mobile phone while he was driving to one of the many consultations he does with farmers every year, teaching them about a system of cultivation that is not just organic, not just sustainable, but that actually renews the health of farmland. It also pulls a great deal of carbon out of the atmosphere while producing food that's considerably higher in nutrients than that produced by the conventional industrialized model. Brown learned the methods the hard way, out of desperation.

In 1995, a few years after he and his wife took over his father-in-law's farm in Bismarck, North Dakota, their entire 1,600-acre crop of grains was wiped out by hail. Such total devastation of a farm's production by hail in the area was unheard of, so the Browns didn't have hail insurance. The next year, in a remarkable fluke, hail again destroyed their entire crop. In 1997, their whole region of North Dakota was hit with a drought so severe that no planting could be done. Then, in 1998, hail again pummeled the Browns' fields, this time killing 80 percent of their yield.

Brown hadn't grown up on a farm—he was a city boy from Bismarck. He'd fallen in love with farming in a class he took in college. But facing financial ruin, he recalls with an edge of his characteristic self-deprecating humor, "I was starting to question my career choice, and my wife was starting to question her choice of husband." Rather than give up, however, he embarked on a journey of discovery that has transformed his farm into one of the world's leading models of regenerative agriculture—a term was coined by Robert Rodale, the son of one of the American pioneers of organic growing, J. I. Rodale. Robert was so inspired by Albert Howard's books that he and his wife bought an old farm in rural Pennsylvania and founded the Rodale Institute in 1947 to further the research the Howards spearheaded.

Brown's insights about the best methods of transitioning a farm from conventional growing to regenerative methods are in such high demand that he was on the road 252 days of 2019. "I get calls daily," he tells me. "Just yesterday I was on the phone with three farmers, all on the verge of bankruptcy." Though regenerative methods have been in development from the beginning of the organic movement, Brown says that now "the snowball is finally starting to roll downhill."

The practices he's honed so well are largely a return to those Albert Howard championed, with the focus on building soil health.

In the ideal version of regenerative cultivation, farmers never till their fields, planting them with a rich diversity of "cover crops": low-growing grasses and legumes, such as cowpea, oilseed, and daikon radishes; the list of possibilities is long. As Brown says in his frequent speeches to farmers, "Where in nature do you find bare soil?" You don't. Cover crops are soil's armor, Brown explains, protecting from erosion, but they also pull water and natural nutrients down into soil, and their root systems intertwine and spread those nutrients around. That allows the plentitude of microbes, as well as earthworms, that should be aerating and enriching soil to thrive. Brown recounts that when he started farming, he couldn't find any worms in his soil, but recently in a sample of soil just one foot by one foot and two inches deep, he counted sixty. He didn't cultivate them. "They came on their own."

By planting a diversity of seven to ten cover crops in each field, the solar energy they convert into carbon soil is optimized. The cover crops also obviate the need for pesticides, because they are great attractors of beneficial insects that are predators of harmful pests. "For every one crop killer," Brown tells me, "there are seventeen hundred beneficial insects." Left to its own devices, nature knows perfectly well how to keep control of its pests.

Seeding of the crops to be harvested is done by pricking holes through the cover crops, with corn or wheat or oats having no trouble whatsoever growing up through them. What's more, the cover crops allow little room for weeds, and regenerative farming uses no herbicides or synthetic fertilizers. The only fertilizers used are either compost or manure, the latter of which is imparted to the soil by grazing livestock out in the fields—where they munch on the cover crops, but eat only about a third of them before they're moved to a new field. Gabe Brown likes to say, "We don't need to provide livestock with a bed and breakfast." They much prefer open grazing and eating the nutrient-rich greens they are built for. And with manure enriching the soil rather than releasing methane, regenerative ranching may be an even better solution to the cattle pollution problem than the booming business of the Beyond Meat and Impossible Burger plant-based meat substitutes. As Mark Hyman highlighted in *Food Fix*, a study, funded by General Mills, of the greenhouse-gas impact of the regenerative ranching practiced on the White Oak Pastures ranch in Georgia, which raises a hundred thousand calves annually, found that the overall impact was net negative gas emissions of −3.5 kilograms per pound of meat; whereas net emissions for conventionally raised beef were +3.3 kilograms per pound, and for the Impossible Burger, a slightly higher +3.5 kilograms per pound.

Brown's results, and those of many other regenerative growers around the world, are persuading farmers in droves to convert to these methods. When he took over the farm, the soil contained only 1.7 percent organic matter, was light tan in color, indicating a lack of nutrients, and retained water poorly. Now, in most fields the quotient of organic matter is up to 6 percent, with one field coming in at 11.1 percent, and the soil is a deep brown. Whereas when he started, the soil could absorb only half an inch of rain per hour, it now soaks in eight inches an hour, a testament to the power of regenerative methods to reduce flooding.

Brown has worked with some of the leading scientists involved in measuring carbon sequestration and the increase in nutrient density in crops achieved with regenerative ag, and here too, the results are profound. While some who study carbon sequestration have doubted that regenerative farming can pull carbon permanently into soil any deeper than two feet, a two-year research project run by soil science firm LandStream, found that Brown's soil has sequestered substantial quantities as far down as eight feet. Brown is also part of the Real Food Campaign, launched by the Bionutrient Food Association, seeking to commercialize the measurement of the nutritional density of food at large scale.

Brown does not participate in any government support programs. He has no need to, because his yields are so high and he has such a diversity of produce to sell. If one or two crops have a bad year, he's got plenty of others to offer. While organic growing has often been said to produce lower yields than the conventional method, Brown's are substantially higher than the average in his county. For example, he produces about 142 bushels of corn an acre annually, while the county average is just under 100 bushels. His yield results are backed up by the longest-running study of the yields of organically grown versus synthetically fertilized fields, started in 1981 by the Rodale Institute, the premier research institution studying organic farming. In side-by-side plots, though yields were slightly lower in the first few years of farming with organic methods, the institute reports that thereafter "the organic system soon rebounded to match or surpass the conventional system." In fact, according to Rodale, switching to organic "can lead to a harvest 180% larger than that produced by conventional methods." That increase in production, combined with no cost for synthetic inputs and the higher prices organic products can command, translates to much greater profitability, with organic growing affording an average net return of $558 per acre annually versus an average of $190 per acre using conventional methods. With data like that now ripping around the farming community, thanks in no small part to Brown's tireless advocacy, it's no wonder the regenerative snowball is rolling.

Additional momentum should pick up after the demonstration of the glaring vulnerabilities of the industrial ag system revealed by the coronavirus. Painful accounts of farmers in Florida plowing under massive, monocultured fields of cabbages, beans, and tomatoes; egg producers destroying hundreds of thousands of eggs; and pig farmers euthanizing entire herds of thousands of animals exposed the excesses and rigidities of Big Ag. Gabe Brown told me that in stark contrast, the organic and regenerative growers he's so constantly in touch with, who sell locally and often direct to consumer, as he does, saw a massive boom in sales as those in their local communities flocked to their farms.

Providing incentive for farms to transition to regenerative methods, a number of investment firms and food companies are offering to finance their transition. Investment firm Steward Partners, for example, created the Steward Farm Trust with a website on which investors can select farms to support. Over two thousand farms had participated by the end of 2019. Building on the work in ecosystems services accounting, innovators Nori and Indigo Ag have both created online marketplaces through which companies seeking to purchase carbon credits can pay farmers for carbon sequestration through regenerative growing. Some of the largest pension funds, such as TIAA, have purchased land and funded its transition to regenerative. As of 2019, $47.5 billion had been invested in furthering regenerative practices in the U.S. alone. Overseas, the UN teamed up with one of the world's largest agricultural lenders, Rabobank, to create a $1 billion fund for assisting farmers in the developing countries to make the switch.

And not far behind is interest from major food producers, which I find particularly hopeful. Dairy products giant Danone is working with dairy farmers to help them move to open grazing of cows on cover crops. Anheuser-Busch launched the Contract for Change program, through which it offers contracts at premium prices over many years to farmers for supplying regeneratively grown barley, rice, and hops. Gabe Brown is helping here too: General Mills hired him to train farmers in North Dakota and Canada in regenerative methods for a multiyear project launched in 2019 to study the results of regeneratively growing 45,000 acres of oats for its breakfast cereals.

Fertilizer and Fuel from Food

The last link in closing the loop of the food system is turning whatever food waste can't be repurposed or redistributed as food—eggshells, fruit peels and pits, discarded chicken skin, and trimmed-off beef and pork fat—back into either soil nutrients or clean energy for growing and cooking food. Here again, India has offered a solution. Yair Teller, one of the founders of HomeBiogas, shared with me the story of how he got the idea for the company when I visited their operations in Israel.

One day, during a hike up a remote mountain on a trip to

India, he stumbled upon a thriving small farm. Welcomed in for a meal by the gracious couple who lived there, he was stunned when he walked into the kitchen and saw the woman was cooking with "this beautiful blue flame." He had seen that the house was entirely off the grid; there were no electricity or gas lines in the area. Yair also knew that most Indians without access to electricity or gas service cook on open wood or coal fires, as do the poor in many developing countries, and such open-fire cooking is both one of the worst greenhouse emitters globally and a killer of millions every year who succumb to smoke-related illnesses.

"I asked, where is this gas coming from?" he recounts, and she brought him to their cowshed to see the simple concrete anaerobic digester they had installed in the floor, a basic system used for thousands of years. They put all their household food waste in it, as well as the manure from their few cows; and tiny microbes in the digester transformed them into biogas, comprised largely of methane and a residue of nutrient-rich liquid fertilizer. Burning biogas converts the methane to carbon dioxide, which, though released to the atmosphere, is twenty-eight times less potent a greenhouse gas than methane. As for the fertilizer, the farm couple used it to grow all their own vegetables, as well as beautiful white calla lilies that they sold in the local market.

Yair was in India in search of a new life direction, having just completed his compulsory service in the Israeli Defense Forces and no longer keen on completing a graduate degree in biology. "I needed to find a purpose," he says, and so entranced was he by the simplicity of the anaerobic process that, he tells me, "Immediately I knew what I had to do with my life." Several years of design and testing later, he and two friends, Oshik Efrati, now the CEO, and Erez Lanzer, launched HomeBiogas, a digester for backyard installation—foldable, about the size of a child's pup tent, and shipped directly to the customer. I was blown away when he demonstrated how it works; it could be a truly transformative solution for health, sanitation, and renewable energy.

For an average household in developed countries, the digester can produce enough gas from food waste for several hours of cooking every day, with an easy hookup to a gas stove. In developing countries, the digesters are saving lives by reducing pollution and diseases that result from the improper management of food and biological waste. By 2020, HomeBiogas has installed thousands of systems in over fifty countries and has gained traction in developed countries with thousands of units sold in the U.S. and Europe. Food waste is converted into gas that can be used for indoor cooking, an outdoor barbeque grill, or a hot-water heater. There is byproduct produced by the digestion process that can be used as a nutrient rich organic garden fertilizer.

Closed Loop Partners provided funding to help HomeBiogas expand their operations and, in a sign of its great potential, French energy giant Engie also invested. Yet another big vote of approval came from the European Union, which awarded HomeBiogas a prestigious Horizon 2020 grant to fund development and distribution of a larger system for food-service providers, including hotels and schools, food companies, and farms, that will significantly reduce greenhouse gas emissions. With widespread adoption, the potential for cutting greenhouse gas emissions and reducing operating costs for food service providers and farms is tremendous.

Yair tells me that selling to schools has been particularly gratifying because it offers him the opportunity to help teach children about anerobic digestion and the food-waste problem. He spends a good deal of his time traveling to schools to assist them in getting their systems up and running, and then showing the children how valuable the food they might just toss is. He's a father of two and he's made sure his own children also learn the lesson—his family lives entirely off the grid in a yurt, and they do all their cooking on a HomeBiogas unit.

Anaerobic digestion is also being adopted at scale by cities, making fuel from their food waste. New York City has been running a massive digester operation in Brooklyn since 2016 that turns 130 tons of food waste into gas annually, and a new facility is being built in Brookhaven, Long Island, that will process another 10,000 to 15,000 tons annually. In Philadelphia, a vacant former oil refinery facility is being transitioned into a $120 million digester operation that will process 1,100 tons of both food waste and manure from regional farms *daily*, with the gas produced used to fuel the city's buses and trucks. Los Angeles and Salt Lake City are also running large anaerobic digester operations. Smaller cities are also utilizing anaerobic digesters. North of Los Angeles, the coastal city of San Luis Obispo, population 47,500, sends all its food waste to a privately run facility and gets back enough gas to fuel six hundred homes.

The other great option for large-scale food-waste transformation is composting, which is the most circular system, turning food waste back into nutrient rich soil. The program we launched in the Bloomberg administration for curbside food-waste pickup in 2013 has grown to include over 10 percent of city residents and the City Council is proposing to expand the service citywide. Most of the collected food waste is processed at local compost facilities and the nutrient rich compost is then sold to local landscapers. Ten percent participation may not sound like much, but it equates to some eight hundred thousand people, more than the total population of most American cities. In San Francisco, which made composting mandatory in 2009 for all businesses and residences, most of the city's food and yard organic waste is turned into soil that's sold to vineyards and farmers in the Central Valley. Vintners effuse about how the "terroir" that grapes grow in produces distinctive flavorings in their wines.

Sir Albert Howard wrote in his magnum opus, *The Soil and Health*, "The world is divided into two hostile camps: at the root of this vast conflict lies the evil of spoliation which has destroyed the moral integrity of our generation." He concluded that "it will not be amiss to draw attention to a forgotten factor which may perhaps help to restore peace and harmony to a tortured world. We must in our future planning pay great attention to food." Though it's taken many decades to gain mass-movement momentum, the legions innovating a circular food economy are now most definitely paying attention.

The Sustainable Closet

W HEN I WENT TO meet with the Renewal Workshop cofounder Nicole Bassett at the company's headquarters
in Cascade Locks, Oregon, I was surprised by how rural
the factory's location is. The town is about an hour east of Portland by car, in the midst of vast expanses of forest, nestled up
against Hood River, with a population of just over a thousand.
Nicole and her cofounder, Jeff Denby, have nonetheless created a
thriving business in the fashion industry purely by tapping into
local talent, working with major brands, and devising a brilliant
model. The Renewal Workshop is a stellar example of how innovators leading a circular fashion revolution are proving that the
industry can dramatically, and profitably, clean up its wasteful
practices.

The Renewal Workshop works with twenty apparel brands, including The North Face, Carhartt, Eagle Creek, Mara Hoffman,
and H&M's high-end Cos line, to refurbish their damaged and
unsold inventory and returned items. Nicole shared with me that
the prevailing practice in the industry is to trash or burn damaged

and returned items—even if they're just missing a button. The brands and retailers don't have the staff to make repairs or clean returns. The waste is astronomical and cost to shareholders significant, which appalled Nicole, who formerly worked as social responsibility manager at Patagonia, among many other industry posts. This product waste and financial loss to shareholders is a result of antiquated manufacturing supply chains that were developed in the 1960s, '70s and '80s to take advantage of ridiculously low labor rates in emerging markets.

At The Renewal Workshop every item is meticulously inspected upon arrival and then washed in machines using Tersus Solutions, a waterless technology that uses liquid CO_2 in a closed-loop system that recovers the liquid for reuse, a cleaning method that's much less harmful to fibers. The clothes are then repaired by highly qualified "sew techs," with any items that are beyond refurbishment to their rigorous quality standards sent for recycling. Most of the refurbished clothes are made available through re-commerce to consumers, such as through The North Face Renewed line.

Why base the business in Cascade Locks? "My favorite moments in life are in the wilderness," Nicole explained, a sentiment her husband shares. When Nicole learned that one of the premier manufacturers of kiteboards was located not far away—Hood River being a windsurfing mecca—she realized she could base an apparel business there too, no offshoring of labor required. She learned reverence for nature growing up in rural British Columbia, out in the wilderness not far from the territory of the Wet'suwet'en people, one of Canada's First Nations. Her mother bought goods from the Wet'suwet'en and took Nicole with her to sweat lodges, where Nicole learned that "the interconnected relationship between their people and the land was filled with honor." She was determined to find a way to reduce the waste of the fashion business because, as she says in a TED talk, "just by getting dressed this morning, all of us damaged the planet."

The same mission to solve fashion's waste problem and unlock the enormous value for companies behind this transition is what drew Caroline Brown to join us at Closed Loop Partners as a managing director. As an industry veteran, with more than two decades of experience including as CEO of Donna Karan, Carolina Herrera, and Akris, she is acutely aware of the opportunities to transform the fashion industry. Caroline grew up in New York City and loved to keep up with all the fashion trends, from multicolored painter's pants to Fiorucci angel logo T-shirts and vintage treasures. What fascinated her about clothes, she says, "was not fashion per se, but more its ability to reflect culture, moments in time, people's values." Once working in the industry, starting her career with a decade at Giorgio Armani, she became intrigued by the complexities of the business. "To make a fashion company work," she says, "so much has to be firing at the same time, and perfectly so, with a great balance of creativity and business discipline." She had a front-row seat as Armani built an empire—which Caroline credits to "extraordinary entrepreneurial vision, discipline, and grit"—transforming his clothing line into one of the first megabrands to offer everything from luxury wear to children's clothing, jeans, home goods, fine dining, and chocolates.

Traveling the world for many years, learning the nitty-gritty of operations, Caroline realized that the industry had an unprecedented opportunity to reinvent itself for the better. Since the advent of industrially produced clothes in the late eighteenth century, the apparel trade has evolved into one of the most environmentally damaging and people-punishing sectors of the global economy.

The horrifying working conditions in so many garment factories have periodically received major media coverage. In recent memory is the collapse of the eight-story Rana Plaza factory complex in Dhaka, Bangladesh, in April 2013, one of the worst factory tragedies in history, killing 1,134 workers and injuring another 2,500, their limbs crushed by falling pillars and disintegrating floors. Of the workers, 80 percent were women in their early twenties, paid on average just over $1.50 a day. The industry has witnessed the improved conditions in many factories in Bangladesh, China, Vietnam, and other hubs, some with state-of-the-art facilities, but too many others continue to perpetuate abuses. And the suffering is not confined to the developing nations. As fashion writer Dana Thomas reports in *Fashionopolis*, her hard-hitting examination of the industry, in 2016 the U.S. Department of Labor cracked down on sweatshops in Los Angeles that hire mostly undocumented immigrants, paying them considerably below minimum wage for work in hazardous conditions.

As for environmental havoc, the industry accounts for an estimated 10 percent of global greenhouse gas emissions. Add to that the 17 to 20 percent of all the world's industrial water pollution, due largely to the toxic fabric dyes still most commonly used. So much water is used for dyeing that one account estimates it equates annually to half the volume of the Mediterranean Sea. Pollution also comes from the fabric itself. With so many of our clothes now made from synthetic fibers, which are various forms of plastic, a third of the accumulation of microplastics throughout the world's waterways is attributed to apparel. These tiny particles, which range from .05 millimeters down to the microscopic 10 nanometers, are about the size of plankton, the protein at the base of the marine food chain, and are gobbled up right along with it. Microplastics have been found to absorb water contaminants, such as PCBs and DDT—still in use all these years after Rachel Carson's harrowing exposé. Researchers have found high concentrations of plastic fiber remnants in fish and shellfish all around the globe, including even the Antarctic. How do those bits get from our wardrobes to the seas? Every time our clothing is swirled and spun in our washing machines, bits of fiber are torn away and drained straight into the water supply. So we are also consuming microplastics in our drinking water. The estimate is that seven hundred thousand microplastic fibers flow out of every typical washing machine load, and that Americans on average eat and drink about seventy thousand microplastic bits every year.

Synthetics aren't the only bane of clean water; cotton is one of the most water- and fertilizer-intensive of all crops to grow. It's also especially vulnerable to insect infestation, resulting in its cultivation accounting for an estimated two hundred thousand tons of pesticides and 8 million tons of fertilizers doused on soils annually. The production of one cotton shirt requires an estimated twenty-seven hundred liters of water, which is enough to meet the daily drinking needs of one person for two and a half years. The loss of water volume in the Aral Sea in Central Asia, and in fact the loss of most of the sea itself, is a particularly stark testament to cotton growing's devasting impact. Uzbekistan has become the world's sixth-largest cotton producer, diverting so much water from the Aral Sea to irrigate land not suited to the crop that the sea, which was once the fourth-largest freshwater lake in the world, is now only 10 percent of its former size. Shocked by the apocalyptic scene on a visit to the sea, UN Secretary General Ban Ki-moon described it as of the worst environmental disasters in the world.

Further contributing to the fashion travesty is the fact that an estimated 73 percent of clothes produced globally end up in landfills, with only 1 percent of fabric recycled, although 95 percent of discards *could* be recycled. In the U.S., the annual haul is calculated at 12.7 million tons, which comes to 70 pounds of fabric trashed per American per year.

Meanwhile, about 20 percent of what's produced never makes it into consumers' hands, going unsold. Most of those clothes are sent to landfills or burned. The total lost value of this "dead inventory" is calculated at $50 billion a year in the U.S. retail industry alone. Why not donate these items to charities? Companies fear their brand will be tarnished if their clothing shows up for sale at Goodwill and the Salvation Army. That's why most excess inventory that is donated is shipped overseas. But even most of those donations end up in landfills or burned also. Oxfam reports that about 70 percent of the clothing shipped by charities overseas goes to sub-Saharan Africa, where wool sweaters are inappropriate to the climate or unsuited to local styles. People in Africa care that they're in style too.

The Ellen MacArthur Foundation conducted a study that found that if these trends of overproduction and dumping continue on course, by 2050 the industry would account for a quarter of the global "carbon budget" annually. Gasp. How did the making of clothes, which for eons was an artisanal craft, with clothing for many being their most highly valued possession, go so far afoul?

The meteoric rise of fast fashion in the past couple of decades has received widespread and much deserved blame, with its five-dollar dresses and ten-dollar jackets, often so flimsy that a seam might suddenly burst open in the middle of a dinner out. But the trend toward throwaway clothing started long ago, with the invention by the E. I. du Pont de Nemours and Company of the first fully synthetic fabric: nylon.

The First Miracle Fiber

DuPont proclaimed the wondrous new fabric was made simply of coal residuals, air, and water, but there was nothing simple, or natural, about its creation. Harvard chemistry professor Wallace H. Carothers was lured to the company to head up a team of 230 scientists who worked for eleven years on the project, discovering a means of artificially stringing together long chains of molecules, called polymers, that could be woven into fabric. DuPont also spared no expense in hyping nylon's revolutionary qualities, putting stockings front and center at its 1939 World's Fair exhibit. The fair featured many lavish futuristic displays, such as Westinghouse's seven-foot-tall talking robot, Elektro, which boasted a seven-hundred-word vocabulary and professed to gawking crowds, "My brain is bigger than yours." But DuPont's nylon exhibit nonetheless snared global media attention. The booth featured seamstresses cranking out nylons, which female models not only wore but played tug of war with to demonstrate the fabric's strength. Nylons were "strong as steel," DuPont claimed.

While some press coverage doubted nylons would catch on,

noting condescendingly that "it's difficult to tell about female psychology," women were thrilled. Advertised as "so durable that they resist runs and even cigarette burns," a limited run of four thousand pair of nylons went on sale in 1939 in six select stores in Wilmington, Delaware, where DuPont is headquartered. Women lined up for blocks and the stock sold out within three hours. The original name of the fabric was nuron, which a manager of DuPont's Nylon Division explained was "no run" written backward. (A trademark for that name owned by another firm forced the change.) One woman reportedly asked a salesperson how many years a pair would last, and some accounts by women of the day attest to the stocking's staying power. Grace Lyons later told a reporter, "They were like iron. They'd last for a year." So why do they run so readily now? Because at some point DuPont reportedly instructed its chemists to find a way to make them less run-resistant. Planned obsolescence strikes again!

Certainly, nylon can be made extraordinarily strong, which is why the U.S. military mandated that DuPont stop making stockings and repurpose all the fabric into producing parachutes and tents during World War II. The resultant stocking shortage led to so-called nylon riots when they went back on sale in 1945, with crowds of ten thousand and more women descending on shopping hubs all around the country. The worst incident occurred when a crowd of forty thousand stocking seekers in Pittsburgh competed for thirteen thousand pairs a small boutique had at last been able to procure. As author Susannah Handley writes in *Nylon: The Story of a Fashion Revolution*, "in all the history of textiles, no other product has enjoyed the immediate, overwhelming public acceptance of DuPont nylon."

The environmental implications of clothing produced from fossil fuel didn't entirely escape notice, with one reporter chastising, again with sexist condescension, "If you're wearing those new nylon stockings, girls, you're carrying around more coal dust than a miner." But who could argue with such success? Other chemical companies leapt into action, concocting a flurry of new synthetics in the decade that followed. The most popular of all was polyester, introduced in 1951 as "a miracle fiber that could be worn 68 straight days without ironing and still look presentable." (Still smelling presentable would, of course, have been another story.) One wonders what strange change would come about on day 69.

Polyester is constituted from petroleum chemicals rather than coal extracts, as are almost all the synthetic fibers created since. They are usually cheaper to manufacture than natural fibers, giving them powerful market advantage. But their wrinkle-resistant convenience was also key to their appeal. As sales of washing machines exploded after World War II, more than tripling in the U.S. between 1950 and 1956, sales of "wash and wear" apparel skyrocketed in step. Requiring even less effort to clean were popular "drip-dry" suits, which, as their purveyors touted, could simply be rinsed in the shower. None other than perhaps the most dapperly dressed man of all time, Cary Grant, demonstrated the feat for Audrey Hepburn in their 1963 caper flick *Charade*, gleefully reading the care label of his suit to a startled Hepburn as he lathers up in the shower: "Wearing the suit during washing helps protect its shape!"

Convenience reached a still-unmatched pinnacle a few years later with the runaway popularity of the first full piece of clothing expressly created for throwing away, the paper dress (paper shirt collars had been introduced back in the 1920s). While touted in the 1950s, the idea didn't take off until 1966. The Scott Paper company had developed a stronger weave paper, and as a promotional gimmick, advertised its sleeveless A-line "Paper Caper," made from the new weave. Available in two mod sixties designs—the geometric black-and-white "Op Art" print and the boldly bright red-and-yellow-flowered "Bandana"—they could be ordered by mail for $1.25, postage included. To Scott's surprise, orders flooded in. Soon, rivals proliferated. The most coveted offerings came from a small firm, Mars of Asheville, in western North Carolina, whose Waste Basket Boutique selections included a floor-length silver number and several candy bar prints, such as the Baby Ruth. A particularly brisk seller was its Yellow Pages print. When an ad for the dress ran in *Parade* magazine with the tagline "What's Black and Yellow and Read All Over?" the company received twenty-five thousand orders that day and another fifty thousand orders the next. Just for safety's sake, the company's care labels warned in all-cap type: DO NOT WASH.

Soon paper pantsuits were the rage; a *Miami Herald* columnist helpfully explained could easily be snipped after a first wearing into a pair of clamdigger pants, then Bermuda shorts, and finally a sexy bikini, warning, though, "Don't swim in it!" But by 1968, the craze had died down, perhaps due to experiences like that of reporter Nancy Hayfield, who recounted that "the first time I wore the paper dress, I was sure it would fall apart. It didn't. The last time I wore it, it did."

While fast fashion can't, therefore, lay claim to the ultimate in throwaway apparel, it's giving the paper dress a hell of a run for its money. Indeed, the $4.99 Sleeveless Jersey Dress sold by H&M is cheaper by about $2.50, in adjusted dollars, than Scott's Paper Caper was. Fast fashion is the culmination of a steady march toward lower and lower costs of production, achieved by "chasing the cheap needle." As clothing brands in the U.S. and Europe moved their manufacturing to Asia in the 1990s, a precipitate decline in clothing prices commenced, even as the costs of most other consumer goods increased. The Consumer Price Index for goods overall has risen 63 percent in the last twenty years, but for apparel has fallen 3.3 percent, which when adjusted for inflation, translates into a 41 percent real decline. While the decline has been much steeper for low-cost brands, the price tags of even many of the most established prestige brands have also fallen considerably. A midrange Brooks Brothers men's suit, for example, would have cost about $600 in the mid-1990s, which, adjusted for inflation, would be the equivalent of about $960 today; yet many were on sale as of mid-2020 for between $300 and $350. One result is that while as of 1990 Americans spent on average between 12 and 14 percent of their income on their wardrobes, it's now down to about 3 percent. That's despite the fact the average consumer is now purchasing 60 percent more clothes annually than in 2000. Meanwhile, the average number of times any given item is worn has plummeted, and, as Elizabeth Cline shares in *Overdressed*, her exposé of the clothing consumption craze, about 70 percent of the average American wardrobe languishes in drawers and closets unworn.

Our hunger for the latest in fashion is well-founded. Most of us are well aware that although the saying goes "We are what we eat," when it comes to how others perceive us, we are what we wear. Psychological studies have revealed many subconscious effects on people's reactions to us caused by our clothes. People who dress similarly to their bosses are reportedly promoted faster, and people more readily respond to a request for money from people who are dressed more like they are. The colors we wear also affect impressions. One study, for example, found that both men and women rate people wearing red as more attractive generally, and research even found that waitresses wearing red T-shirts received higher tips from men, but not from women, than waitresses wearing a range of other colors.

Our clothes also affect how we feel about ourselves and how we think. Psychologists have studied what they call "enclothed cognition," which is the impact of what we wear on our self-assessments, our moods, and our interactions with others. Wearing tailored suits, for example, has been shown to put people in a more focused, analytical frame of mind. One study showed that donning a white lab coat also boosted cognitive performance, with those who wore them making fewer mistakes on a set of tasks versus another group who wore casual street clothes.

The good news is that attitudes about what our clothes say about us have been rapidly morphing. The revolution of stylish, cheap apparel was so successful because it was democratizing, allowing so many for whom dressing in the latest looks had been prohibitive to raise their style profile. That was much to be ap-plauded. But as the horrifying revenge effects have been exposed, the messaging we broadcast by wearing cheap clothing has rapidly turned. Millennials in particular are showing a strong commit-ment to buying sustainable clothes. In survey after survey, they've also reported that they are happy to spend more on clothing that's sustainably made. Their self-reporting is backed up by a remark-able trend in Google searching that Caroline Brown showed me. Beginning about 2014, searches for "cheap clothing" began to de-cline precipitately, while at the same time searches for "sustainable clothing" spiked. The lines crossed in 2017 and searches for sus-tainables have soared since.

Those same years have seen an explosion of attention to sus-tainability in the industry, following in the footsteps of sustainable fashion pioneers such as Stella McCartney, Eileen Fisher, and Pat-agonia's Yvon Chouinard. We're seeing innovation driven both by lean and scrappy start-ups, like Fashion for Good, an innovation accelerator that assists other start-up founders to scale up their solu-tions, supported with funding from Adidas, Kering, PVH, Chan-nel, Target, and Stella McCartney, among others. Global nonprofit Fashion Revolution, based in the UK with hubs in one hundred countries, holds its Fashion Revolution Week every year to mark the anniversary of the Rana Plaza factory disaster. They host events to raise awareness, such as clothing swaps, and keep up a steady stream of messaging through a number of popular Twitter hashtags such as #WhatsInMyClothes.

Especially promising is the phenomenal commercial success of innovative business models built on circular principles. The ap-peal of one model, recommerce, has been resoundingly proven by Rent the Runway. Several rivals, such as Gwynnie Bee, Le Tote, and Haverdash are building strong followings, and many major brands now offer rental services, including Ann Taylor, Urban Outfitters, and Banana Republic. The potential for waste elimina-tion as the rental market continues to grow is immense. Not only is the typical rental item worn thirty times versus the few times so many owned items are worn, but the rental firms are strongly incentivized to purchase durable clothes.

Brilliant innovation is flourishing all around the full apparel loop, from prefabric to end-of-use. A 2020 Boston Consulting Group report concluded, "A perfect storm of innovation and opportunity is forming in fashion." As Caroline says, "Great transformations happen at critical moments, and this is one of those for fashion." She sees a profound shift under way akin to the disruption of the music industry, and following that with consumers' embrace of organic and locally grown food.

Ecosystems Accounting Comes to Fashion

Few in the industry can claim to have done as much to inspire change as former Puma CEO Jochen Zeitz. When Zeitz took the reins at Puma at thirty, he was the youngest CEO in German history, entrusted with the fate of one of the country's most storied brands—and the future didn't look good. In 1993, the year Zeitz took charge, despite an illustrious history of top athletes sporting Pumas, from soccer legend Pelé to basketball icon Walt Frazier and tennis great Martina Navratilova, the company was facing bankruptcy as a low-cost has-been. Zeitz recalls that when he joined, the staff felt "as if failure and negative thinking were clinging to the walls of our buildings." But he had served for three years as Puma's marketing director, and he had a visionary plan: turn Puma into a fashionable "sports lifestyle" brand. He created a whole new category of apparel, now usually called athleisure. At an estimated market size of $155 billion in 2019 and still growing rapidly, it's one of the most successful innovations in fashion history.

Zeitz is not one to be timid. To make Pumas the sexy "sneaker for the street," as he put it, he commissioned brashly colored and flamboyant styles from edgy designers such as Alexander McQueen. Celebrities pounced. When he signed sprinter Usain Bolt, now "the fastest man in the world," to a $1.5 million endorsement contract in 2003, Bolt was an unknown. Zeitz featured him in a global advertising campaign. Five years later, Bolt won gold in the 100 and 200 meters at the Beijing Olympics and exuberantly kissed his golden-colored Pumas for all the world to see.

He wasn't about to take baby steps in addressing Puma's environmental footprint either. He commissioned Puma's first EP&L assessment because, as he said to me, "I didn't want to be someone who says, 'Look! We have a solar panel on the roof.'" An environmental profit and loss statement provides transparency into the profit or loss that brands' manufacturing and sales provides society. He went to considerable expense to get the assessment crafted, and then publicly announced its sobering findings in 2011, estimating Puma's total environmental impact to society at 145 million euros annually. He then posted the detailed results online for anyone to explore. In doing so, he set a transformative new standard for transparency, way ahead of consumer demand. As Caroline recalls its effect on the industry, "For a company to invest that money now, you can see a clear rationale, but back then, it was very brave. It set a great precedent."

Zeitz clearly saw, he told me, that "sustainability is not only a responsibility, it's an opportunity, to retain the best staff and to show customers that the company is doing more than just serving shareholder value." Though in that regard, Zeitz had no worries. Puma's share price had risen from $10.86 when he took charge to $442 by the time he decided in 2011 to become the head of sustainability for Puma's largest shareholder, the massive global fashion holding company Kering. In that role, he promptly led the creation of an EP&L for Kering overall, which is also available for viewing online. The group's EP&L showed a 77 percent reduction in greenhouse gas emissions between 2015 and 2018.

A flurry of related assessment tools have now been created to allow apparel brands to measure and report on their environmental impact and target the best opportunities to become more sustainable. One is the Higg Index, developed by the Sustainable Apparel Coalition. An especially appealing transparency innovation, I think, is the use of a QR code label by sustainable luxury fashion brand Another Tomorrow, founded by Vanessa Barboni Hallik. Scanning the code will tell you where the organic cotton or ethically produced wool an article is made from was sourced. This goes some way toward an idea Zeitz told me he'd love to see implemented: a full environmental health label for every garment, along the lines of the nutrition information provided on almost all foods now, detailing the contribution to greenhouse gas emissions, water pollution. In 2020, Zeitz was named the CEO of Harley-Davidson after previously advising the company on the development of its first electric motorcycle. He will now have the opportunity to share his vision and management expertise with another industry.

A big revelation of the Puma EP&L was that 57 percent of the environmental harm the company caused was due to the production of the raw materials it purchased. The lion's share of that was methane emissions from the cattle raised to provide shoe leather. With awareness such as that spreading about the industry's pre-fabric impact, one of the most dynamic areas of innovation is in developing a host of new, naturally, sustainably, and even regeneratively grown fabrics. The EP&L was an effective way to identify and eliminate waste and inefficiencies in their supply chains.

Who Needs Cotton When There's Algae?

What do pineapple stems, banana peels, mushroom roots, coffee grounds, and milk have in common? They are all among materials now being turned into fabric, driving a "vegan fabric" boom. Start-up Bolt Threads, run by chemistry PhD Dan Widmaier, created a leather replacement called Mylo from the mycelium of mushrooms. Stella McCartney uses the material for her line of Falabella handbags. Created out of waste pineapple leaves, Piñatex is another leather stand-in, which has been used by H&M and

Hugo Boss. Proving the maxim everything old is new again, some of these fabrics are not new but rediscovered. Soybean cashmere, made from soybean waste, is a new version of fabric created under the direction of Henry Ford in 1937 for upholstering Ford cars. Banana fiber textiles, made from waste peels, stems, and bark, were popular for centuries in Japan before they were displaced by cotton and synthetics.

The source of new fabric I'm most excited about is a tiny plant that can be grown in abundance with limited energy and water requirements: microalgae. Daughter and father team Renana and Oded Krebs are among a number of innovators developing fabrics derived from algae. Their ambition is bold. "We aim to be the green engine of the fashion revolution," Renana told one reporter. When I read that, I knew I had to meet her and her dad, and I journeyed to Israel's Negev desert to see the closed-loop cultivation system in action.

The Negev is a fitting, if improbable, setting for visionary innovation. In 1948, after he became prime minister of the new state of Israel, David Ben-Gurion declared, "It is in the Negev where the creativity and pioneering vigor of Israel shall be tested." Indeed, both creativity and vigor are needed in spades, because the Negev is one of the most formidable environments on Earth. A dune-swept moonscape, it borders the Dead Sea, features no rivers or lakes, has less than an inch of rainfall a year, and its temperatures soar to a scorching 120 degrees in summer and dive below freezing in winter. It now houses a major city, university, kibbutz, a number of nature preserves, and innovative companies. If the Negev can become a hub of sustainable algae cultivation, then the prospects for fully circular fabric should be a reality within the next few years.

Oded and Renana have the perfect skills for building Algaeing. She is an award-winning fashion designer with fifteen years' experience working in the industry, and Oded is a plant physiologist who has traveled the world working with energy companies to develop plant-based biofuel alternatives to fossil fuels—algae fuel being a prime candidate. But their passion for helping catalyze a revolution in the apparel industry, and Renana's dogged determination in realizing her vision, are what most impressed me.

Renana never expected to become an entrepreneur. She'd made a splash in the international fashion world while still an undergraduate at Israel's Shenkar College of Engineering, Design and Art. Her final project was a line of men's suits and briefcases with moss growing on them, titled "Greenhouse Effect." She spent six months in a lab working to find the right breed of moss that would continue to grow once "planted" in a thick linen fabric. Her purpose was to call attention to the connection between our clothing and the natural world, and she chose moss because, as she shared with me, "it's an incredible plant" that can live without water for twenty to thirty years and then immediately be revitalized.

Renana wanted to create clothing "that will still have its own life after it has been made," she explains, while also calling attention to the devastating effects of the apparel industry on the planet. She developed a love of plants, and moss specifically, as a child, watering the bonsai trees her father grew in greenhouses as a side business on the family's small farm in northern Israel. As is traditional with bonsai trees, Oded grew moss around their bases. He is also a nature lover, whose career has been about learning from the symbiosis of the natural world. "When you are working every day with plants," he tells me, "you see what nature has to give you from plant systems."

Renana was told she'd never be able to get moss to grow on her clothes, but she was resolute—and her hard work paid off in international headlines when she showed the line at a design award show, one reporter quipping, "Renana Krebs does not wash her clothes, she waters them." She went through months of negotiation to be able to show the clothes because the border control of New Zealand, where the show was held, wanted to restrict them as organic matter. That tenaciousness has been vital in building Algaeing.

By the time of the show, she had worked as a designer for many years for a company in southern Germany and traveled to the fabric-making centers of Asia, witnessing what she calls "the modern slavery" of the industry and the despoliation of rivers turned crimson, purple, or orange by toxic dyes. For her final project in graduate school, she created a flowing white algae fabric; the strong industry interest in the fabric inspired her to start Algaeing, and in 2016 she won the prize of representing Israel in the Creative Business Cup Challenge, the "world championship for creative entrepreneurs," and was a top-five finalist. Next, she won a spot in an accelerator program run by Fashion for Good, and in 2019 she was one of four entrepreneurs selected for its follow-up scaling program. Algaeing also won a coveted Global Change Award in 2018, which came with 150,000 euros, awarded annually by the H&M Foundation to support the best advances in fashion sustainability.

All that recognition is due not only to Renana's considerable talent as a designer and brand builder, but to the potentially transformative potential of the fabric-making technology she and Oded have developed. They combine microalgae with wood pulp, entirely sustainably grown, and then extrude the mixture through a fine mesh to make fibers. She chose algae as their primary ingredient not only because it can be grown in a way that's healthy for the environment, but because, as she says, "it also grows remarkably fast." After scouring far and wide for suppliers, they chose to source their algae from a producer in the Negev because of the remarkable innovations in environment-friendly agriculture that have made the desert a world leader in sustainable growing, such as the drip irrigation system, which delivers precisely calibrated drops of water to crops through thin piping. Its incredible efficiency of water use has made it a vital tool for growers all over the world who are coping with the intensifying water scarcity crisis brought on by climate change.

The Negev is a world-leading cultivator of algae, in a highly competitive industry that is expected to grow to $5.38 billion globally by 2025. As I looked out with Renana and her father at acre upon acre of beautifully crimson and sapphire algae "fields"—which are really long rows of piping on which the algae grows—I recalled the beauty of the riotously colorful tulip fields of Holland, stretching like rainbows rising out of the earth as far as the eye can see. I'm sure the Dutch, who know a thing or two about water engineering, would be impressed.

The circularity of the growing system is vital to the Krebs. The piping optimally conserves the water, not only recirculating it but constantly filtering it, and the algae is fed no pesticides, because the water has been purified of all pests. The combination of sustainability and the potential for very high-volume production give the Krebs faith that algae fabric-making can scale up to mass production. Indications that Algaeing may well lead the way are good; the company is currently running pilot tests with major brands for a line of activewear and a line of bedding and towels.

Perhaps, as Renana hopes, cotton's days are numbered. But in the meantime, some exciting work is going on with regenerative cotton growing. Patagonia has worked with a number of cotton growers in India to help them transition to the method. Eileen Fisher is sourcing wool from regenerative farmers, and in the company's Horizon 2030 manifesto praises regenerative growing as a key component of its "Choosing Circles over Lines" vision for the industry's future.

Using planet-healthy materials for our clothes is only the starting point, of course. Making sure they're worn longer and recycled is also vital.

Reusing and Renewing

One innovator expanding recommerce in fashion is fast-growing used-clothing purveyor ThredUp, which sells brands online from Lululemon to Coach, Kate Spade, and many others, 20 to 90 percent off store prices. In a game-changing proof of concept, Walmart went into partnership with ThredUp in 2020. Another model Caroline and I were particularly impressed by is that of Thrilling, whose website is ShopThrilling.com. Founded by Shilla Kim-Parker, who previously worked in investment banking and at Disney ABC Television, it's a reseller of vintage clothing. She got the idea for the company after the birth of her baby left her with no time to peruse her favorite vintage boutiques in her hometown of Los Angeles. She's using technology to break vintage out of local market constraints. Thrilling collects items sent in by vintage stores all around the U.S. and brings them to their own studios for photographing and listing on the site. The items are then returned to the stores, and if they're bought through Thrilling's website, the company takes a commission. If they're bought at the brick-and-mortar store, Thrilling takes no proceeds. By making vintage shopping so much faster and offering a larger inventory, Shilla is helping revitalize a retail sector that's really been struggling.

Another area of great activity is the creation of new methods of clothing recycling. Until now, apparel recycling has been stymied because so many clothes are made from fabric blends, and teasing the different types of fiber apart has proven either impossible or economically impractical. The mechanical process used, which pummels fabric much as paper is beaten back into a pulp, also degrades the quality of the fibers. The term chemical recycling is now being used to refer to an exciting new process that can break fibers down to their basic chemical components that then become a circular source for new fibers.

UK-based Worn Again Technologies is one leader in the emerging space that has attracted funding from H&M among many others. That's vital, because as Worn Again's chief scientific officer Adam Walker explains, "Chemical recycling only makes sense if you've got enough throughput through your plant to be able to generate really large quantities—five-figure-ton quantities." With keen interest from H&M, which has been accepting clothing in its stores for recycling for several years, along with many other major brands, large quantities of supply won't be a problem. H&M has become so focused on extracting itself from the downward spiral of fast fashion that it worked with the Hong Kong Research

Institute of Textiles and Apparel to help develop a chemical technology dubbed Green Machine that can recycle pure polyester.

As the circular fashion revolution marches forward, inventive solutions that combine new capabilities to create ever more closed loops of production are emerging. One of the most impressive of these is the fully circular model of For Days, launched in 2018 by industry veteran Kristy Caylor. For Days doesn't have customers, it has members. The company launched with only one basic wardrobe item: 100 percent recycled organic cotton T-shirts. That may sound a bit mundane, but Kristy's crafting of the business was anything but.

She began her fashion career working for the Gap, helping launch a number of businesses, which eventually took her to Japan. She had an epiphany when she saw the scale of the company's manufacturing there, realizing, as she told *Forbes*, "We were just making so much stuff." She also realized that we have no connection with the people who are making our clothes, and she wanted to change that. She told me about another eye-opening experience, while on a trip to visit factories in China, that fueled her motivation. The workers at a large factory she toured lived in a modern-day company town built nearby, but the factory and housing facility itself was so large that it was more a full-blown city than a town. The magnitude of the industry's impact, not only on people but the planet, sunk in. When she returned to the U.S. from Japan, she embarked on her first company, the artisanal luxury goods brand Maiyet.

Traveling all around the world to develop relationships with artisans using sustainable materials and methods, she found herself profoundly moved by their commitment to their crafts. In Indonesia, she was struck by the dedication of women carving beautifully intricate items in tea in a tiny village with only one lightbulb and houses with dirt floors. In a village just outside of Nairobi, Kenya, she met a husband-and-wife team making gorgeous jewelry from reclaimed bronze using handmade sand and sugar molds in their tiny backyard. By selling their work through Maiyet she was able to help them move into a proper shop. Maiyet pioneered bringing luxury status to such high-quality craft items, selling them to elite stores including Barneys, Neiman Marcus, and Saks Fifth Avenue.

Kristy first learned about the circular economy concept when she was invited to participate in the Fashion Positive initiative, spearheaded by the Cradle to Cradle Products Innovation Institute in 2014. It has brought fashion industry leaders together, including Eileen Fisher, Stella McCartney, Banana Republic, and Athleta, to establish standards for circularity in the industry. "The potential of circularity was thrilling," Kristy told me, and she decided she wanted to create a truly complete circular model for a new company—and leave luxury items behind. Why start with T-shirts? Because people who wear T-shirts tend to buy, and throw away, lots of them. On average, Americans buy ten a year and throw away six of them, and many avid athletes and casual clotheshounds buy considerably more. If she could offer them as rentals, she could get them back at end of their life and recycle them, using them to weave new fabric to make more shirts. A beautiful closed loop.

But T-shirts aren't well suited for the rental model, because, for one thing, they're so cheap, many costing less than a Starbucks coffee, but also because people aren't keen about wearing someone else's once sweaty workout item. So Kristy landed on a model whereby members pay a basic, one-time fee of $38 to create an initial trial kit of five items, which they purchase at full price, delivered in a returnable bag. They can then swap items for new ones whenever they decide they're done with them. Shirts can be returned in any condition, because For Days recycles them. What's the value add for members? Well, consider that millennial households spend on average $347 a year on T-shirts. With the For Days model, they can spend considerably less for the same number of shirts, but they'll be getting brand new ones regularly. I'd venture that freeing up a good deal of closet space is another plus. Offering a great customer experience and building a strong ongoing relationship with members are fundamental to the model—which,

as Caroline Brown highlights, for fashion brands is a great differ-
entiator.

So successful has For Days been, especially with T-shirt buying millennials, that Kristy was able to build the company its own
manufacturing facility in Hawthorne, California, just outside of
L.A. Assuring recycled material would be "cost neutral," she tells
me, was core to the whole model, for it to be circular. Kristy also
invested up front in a sophisticated inventory-tracking system that
allows her to see exactly what's in each For Days member's "closet"—
meaning all the items they've currently got—and that records all
swaps made, so she can fine-tune her production of new items, solving the overproduction problem.

The potential for the model is profound, given that, according
to the results of a study described in the *Harvard Business Review*,
83 percent of apparel shopping journeys are made for repeat purchases of staples of people's wardrobes. Already, Kristy has expanded For Days' offerings to include sweat shirts and pants and
dresses. She plans baby and children's clothes as her next frontier,
and she's also looking to add wool items.

With so much innovation proving the superiority of circularity for fashion, more and more industry leaders are rallying behind the cause. Results like those Kering has reported through
Zeitz's EP&L are setting the pace. Zara released its first sustainability plan in 2019, with the goal of eliminating hazardous chemicals throughout its supply chain. In 2019, former CEO of Unilever,
Paul Polman, in his new capacity as founder and CEO of the
IMAGINE Foundation, managed to convince the CEOs of fifty-
six leading fashion companies to sign The Fashion Pact, a pledge
to take "all measures" to get to net-zero impact by 2050.

When I asked Jochen Zeitz what it will take for all the large
firms to truly scale circularity companywide, rather than just creating marginal sustainability offerings, he said, "It has to come
from the CEO; if the CEO doesn't push it personally, it won't happen." Which makes Polman's achievement all the more heartening, along with Caroline Brown's observation that "it's the CEO's

job to align with consumer values, and we now have an expressed consumer value set." The values imperative was pronounced by none other than industry thought leader *Vogue* in its January 2020 issue, which declared that the key word for the issue was values: that "fashion needs to reassess its value system, and quickly," urging also that consumers "have to shop with brands whose values reflect our own."

7

I've Got One Word for You, Benjamin

IT'S NOT EASY TO get to the bottom of the Mariana Trench, the 1,580-mile-long, 43-mile-wide ditch that cuts through the Pacific Ocean. At its lowest explored point, nearly 7 miles below sea level, the trench is pitch black, with atmospheric pressures over a thousand times greater than we enjoy on the Earth's surface. To visit, you need a special class of submarines built to handle these incredibly hostile conditions, ensuring that fewer humans have explored the trench than have stood on the surface of the moon. One of those intrepid adventurers, Victor Vescovo, made a shocking discovery in May 2019. As he was filming the murky depths of the trench, his camera caught sight of a plastic bag listlessly wafting its way along the seafloor. Researchers of the ocean plastic problem had previously thought bags couldn't sink anywhere near so deep. If a bag had descended to the planet's deepest depth, just imagine, they realized, how many more of them are lurking throughout ocean waters. Shortly thereafter, research revealed that in some parts of the trench the amount of plastic paraphernalia piled up exceeds levels in some of China's most polluted rivers.

In total, 8 million tons of plastic end up in our oceans each year, in the form of not only the expected bottles, bags, foam clamshells, and straws, but also the little pellets of raw plastic, called nurdles, about the size of a grain of rice, that are the basic ingredient for mixing the panoply of different plastic brews. By some counts, the single most voluminous type of ocean plastic is discarded fishing nets. And then, of course, there are the micro-plastics from our clothes, from beauty products, and from the breakdown of larger plastic refuse.

Much of the debris is swept up into circular ocean currents called gyres, the most infamous one now referred to as the Great Pacific Garbage Patch, estimated as roughly the size of Texas. While we might envision the patch as a solid, floating island, in fact the debris is strewn about into what one researcher describes as a plastic smog. We can think of these massive accumulations of plastic as worse than petrochemical spills because unlike an oil spill, centuries will likely pass before they are broken down and assimilated by nature.

The recent consciousness raising, and public outrage, about ocean plastic pollution should galvanize us to tackle the larger problem of how plastic is managed. Of the 300 million tons of plastic produced every year, less than 9 percent is recycled, with the remainder mostly ending up in landfills or our oceans and rivers. In the U.S. alone, nearly 28 million tons are deposited in landfills annually, and that figure is growing. We might think it's been safely shunted away, but even while the ultimate biodegradation of plastics is excruciatingly slow, all along the way, they leach toxins that make their way from landfills into the soil and water. Additionally, plastics account for an ever-growing portion of fossil fuel extraction, propping up the profitability of the petrochemical firms.

It is important to recognize that plastic has its benefits. It is lightweight and can be molded into different forms. The biggest challenge it presents is that in its original form, it is derived from oil—meaning that the making of it can release a significant amount of greenhouse gases. Therefore, not recycling it means additional extraction and release of harmful greenhouse gases in the production of the next virgin plastic. Even worse, if it is deposited in rivers or oceans, it slowly degrades into its original oil-based chemical form, for us to digest in the fish we consume. How did the plastic waste problem get so out of control? Research we conducted at the Closed Loop Partners' Center for the Circular Economy showed that there is market demand totaling $120 billion for recycled plastic feedstock, from major brands such as P&G and Unilever, as well as from the soda behemoths too. Yet only 6 percent of that demand is being filled. There are enormous investment opportunities to fill this gap by investing in plastic recycling infrastructure.

The plastic waste problem developed because an incredibly multipurpose material with so many advantages for making so many products, not least that it was a much lower cost than materials it replaced, like steel, went from being a durable replacement for those higher-cost materials to being crafted just to be thrown away. There was nothing inevitable about plastic becoming such a scourge, and one reason it's been so hard to eradicate even the most reviled forms of plastic is that, as historian Jeffrey L. Meikle wrote in *American Plastic: A Cultural History*, "It is hard to do justice to plastic because it serves so many functions, assumes so many guises, satisfies so many desires."

From Magical Wonder to Curse

There are naturally forming plastics in nature that are not fossil fuel based. Cellulose, the material from trees used in paper, is one, its strength helping trees grow tall. It's also the core component of cellophane, the first plastic food wrap, which was biodegradable, making its overthrow by the nondegradable Saran Wrap so unfortunate. Another natural plastic, keratin, appears in animal horns, and its light weight and translucence when sliced thinly made it a favored material for the panes of lanterns in the Middle Ages.

The earliest successfully commercialized synthetic plastic, celluloid, was crafted as a replacement for natural materials from endangered species. Invented in 1869, it was created as a stand-in for ivory billiard balls because elephants had been so ruthlessly hunted that obtaining ivory was becoming difficult. It was also used to make fake tortoiseshell, the natural form of which comes from the shell of the hawksbill sea turtle. The beautiful material was in high demand for use as a decorative material for jewelry, furniture, and especially for a nineteenth-century comb craze. As Susan Freinkel recounts in her book *Plastic: A Toxic Love Story*, wealthy women in Europe and America at the time were growing their hair to extraordinary lengths and arranging it into elaborate sculptures of extraordinary height. Tortoiseshell combs were the preferred means of holding the extravagant coiffeurs together.

So voracious was demand that the hawksbill was soon hunted to near extinction. Its scarcity even led men to kill for it in the Ngatik massacre in 1837. Captain C. H. Hart of the Australian trading ship *Lampton* and his crew spotted what they believed was a rich trove of hawksbill shells on the Pacific atoll of Sapwuahfik and returned, armed for battle, to seize it. After murdering as many as fifty natives, they found a mere twenty-five pounds of hawksbill shell amid a much larger cache of worthless (to them) green sea turtle shells. Though an investigation into the incident was launched, Hart and his men were never charged.

Though imitation tortoiseshell was invented in the 1880s, it has not saved the hawksbill. They are still critically endangered. That's in part because they are prone to eat plastic bags floating in oceans, which they mistake for their usual meal, jellyfish.

Such sordid outcomes of synthetic plastic creation were unanticipated. Plastic was seen as a marvelous "material of a thousand uses," and a steady profusion of new plastic products appeared in the early decades of the twentieth century: cigarette holders and ash trays, radios, telephones, cameras, toothbrushes, buttons, and plates and bowls. Scotch tape was introduced in 1930 and Saran Wrap in 1933. But it wasn't until World War II that plastic production really boomed, as metals were devoted almost entirely to making ships, tanks, planes, and munitions. The U.S. military was forced to experiment with materials to replace them, and during the war plastics were used for everything from airplane cockpits and body armor to helmet liners and parachute cords. The government massively subsidized the ramping up of production facilities, and as Susan Freinkel reports, the production of plastics shot up during the war to 818 million pounds in 1945 from 213 million pounds in 1939. American manufacturers prepared for a postwar plastic revolution, joining together in 1937 to form the Society of the Plastics Industry (SPI). The group launched a huge promotional campaign even before war's end, planting newspaper articles about plastic wonders soon to come, with headlines like WHAT'S BEHIND THE BOOM IN PLASTICS? MANY AMAZING WAR USES TO FIND PERMANENT PLACE IN INDUSTRY.

The industry's first big public display of its postwar splendors was a huge success, attracting so many more public visitors than anticipated that it almost had to be shut down due to safety concerns. SPI held the first National Plastic Exposition in New York City in 1946, at which the new wares were ogled by more than eighty-seven thousand visitors. Ever at the inventive forefront, DuPont introduced the fabulous nonstick Teflon. The aptly named exhibitor Billy Glass had turned another new plastic, plexiglass, into musical instruments, from violins to trumpets and snare drums. Booths boasted window frames that would never need painting, suitcases that were super strong yet lightweight, stain-proof plastic upholstery, plastic coat hangers, shower curtains, tablecloths, and shoes. A newspaper reporter summed up the bounty, writing that one could find everything from "darling little cribs to burial caskets molded from plastics." The show's organizer, Ronald Kinnear, crowed, "Who would have thought a brief twelve months ago that it was possible to mold a motorboat?"

The superiority of plastics for many uses was undeniable. Tupperware, also introduced in 1946, kept food fresher so much longer and became the occasion for wildly successful Tupperware parties. So beloved was clingy plastic wrap that Susan Freinkel writes, "People were so enthralled with plastic . . . that the word 'cellophane' was designated the third most beautiful word in the English language, right behind 'mother' and 'memory.'"

A Scripps-Howard article asserted in 1947, "No golden promises of postwar development have come closer to reality than those of the plastics industry." Trouble was on the way, though, as single-use plastic creations proliferated. Most of the initial plastic products weren't intended for fast disposal; in fact, durability was one of the qualities most emphasized about plastic. But then, in the 1950s, the business of plastic throwaways boomed, as celebrated so gleefully in *Life's* "Throwaway Living" photo. The first throwaway plastic bags were introduced, not for groceries but for garbage and for wrapping dry-cleaned items. The advent of fast-food chains spurred the rise of the throwaway foam container market. One result was that already by the end of that decade, scientists discovered that sea turtles and other sea animals were eating plastic. But the findings were buried in academic papers. It wasn't until a decade later that the first findings about the extent to which plastics were clogging oceans were made, and the SPI was immediately on the case with intimidation tactics to tamp down concern.

As reported by journalist Tik Root, oceanographer Edward J. Carpenter was the first to spot the proliferation of plastic out at sea, in 1971, while conducting research for the Woods Hole Oceanographic Institution in the North Atlantic. Shortly after, while doing research along the New England coast, he again discovered a high concentration. After he published his findings in two articles in the prestigious journal *Science* in 1972, SPI sent personnel to Woods Hole to grill him in front of his boss. He told Root, "It was obvious that they were pretty upset about it"; he found their questioning "kind of intimidating."

A year later, in 1973, plastic entered bold new terrain when engineer Nathaniel Wyeth, of the famed Wyeth family, invented the PET bottle, which was the first plastic bottle that would hold up to the pressure of carbonized soda. A plastic bottle that could contain the fizzling and popping of the world's favorite drink, was so much cheaper than glass to make, wouldn't break if dropped or jostled in transit, and was so lightweight that it would cost so much less to transport was such a boon for drink makers that they still fight furiously against calls to abandon it. At least it's not truly, technically, unbreakable; it can be pulverized through mechanical recycling into plastic bits used to make new bottles and various kinds of packaging. PET bottles can also be easily sorted out of single-stream mixes. Which is why PET is the most desirable of plastics for recycling—but even so, only an estimated 30 percent are being recycled in the U.S. Why is that? I'm asked all the time. It's because the country failed to build up the infrastructure to do the job well. For that, the plastics producers and their clients share a great deal of blame.

Why We Haven't Been Recycling

Jeffrey Meikle, who extolled the many virtues of plastic, counted as one of them who said it "so quickly recedes into relative invisibility." Anyone walking around a park or along a city street in the U.S. in the 1970s would have wondered what on earth he was talking about. Plastic waste was on audacious display. In the famous Crying Indian ad, much of the waste the camera panned over was plastic. Once the scourge of throwaway plastic became so visible, calls to do away with it ensued. Progress in other measures to curtail waste was slow, but some impressive victories were achieved. By the 1980s, public calls for bans on foam packaging rose to such a crescendo, with particular fury aimed at McDonald's, that in 1990 the chain relented and announced it would abandon the use of it. In 1988, Suffolk County, New York, passed a ban on plastic bags, and bag bans were championed thereafter all around the country. Calls to beef up plastic recycling also mounted.

As revealed in a powerful 2020 episode of the PBS show *Frontline*, to stem the tide of demand for reform, a number of plastics manufacturers engaged in a devious stunt—supposedly extolling the advent of large-scale plastic recycling. Lew Freeman, a vice president of the renamed SPI, now called the Society of Plastics, was summoned to the DuPont headquarters in Wilmington. An executive told Freeman, "If we had $5 million, we could solve this problem." What was the plan? Straight out of the Disinformation Playbook, they first created an impressively deceptive name. They chose the Council for Solid Waste Solutions. This council comprised representatives of Exxon, Chevron, Dow, and many other of the largest plastics producers. They hired an industry insider, Ronald Liesemer, to be the guy, as he saw his mandate, "who made recycling happen" in the U.S. He had a multimillion-dollar budget, but tellingly, no staff. What did the money go to? Mostly advertising in praise of plastic, with such catchy slogans as "Glass? That's the past."

But some of it went to funding what was pitched to the press as a "million-dollar plastic-sorting system" sent in 1994 to a recycling facility in Oregon called Garten Services. News footage showed the remarkable plastic-sorting process in action. Yet, within a few years, *Frontline* discovered, the sorting machine was shut down and sold as scrap. No serious efforts were made by the council to develop sorting infrastructure, or better collection either. The head of the SPI, to whom Liesemer reported, admitted to the *Frontline* reporter on the phone: "I was the front man for the plastics industry, no getting around it."

The worst irony of this story is that the industry had concluded at that time that plastic recycling would never become a viable business, because, as argued in industry documents *Frontline* dug up, it determined "there are no effective market mechanisms for mixed plastic." That's apparently why it felt secure showing the public footage of sorting happening. With no market for the sorted goods the machine sorted out from one another, what recycling facilities would invest in such equipment? The plastics industry had nothing to fear. Just think of the lost opportunity; after all, if the producers had truly wanted to "make recycling happen," who better to have created not only a viable but a vast market? Unfortunately, before long, a vast market did open up, in China. Starved for stock for its own booming plastics industry, in 1992 China put out a call to all the world to send its plastic castaways, unsorted just fine, and paid good money for them. Thereafter, until the announcement of National Sword, it imported 45 percent of global plastic waste.

Now that the Chinese market has dried up, the domestic plastic recyclers in the U.S. and Europe have a golden opportunity to reinvent their industry for sustainable growth. The good news on that front, as we at Closed Loop Partners have found in our research, is that a tremendous amount of innovation in new plastics recycling processes has been under way in recent years, in labs and recycling facilities all around the U.S. and the world.

But before I share some details of those advances, let me address the number one question people ask me: *Which plastics can and can't be recycled?* From a technical standpoint all plastics can be recycled. But the only plastics that will get recycled into new products are the ones that have robust recycling markets and are profitable to recycle. Those include PET (e.g., beverage containers), HDPE (e.g., laundry detergent containers), and rigid polypropylene (e.g., bottle caps). They're all in one of two basic categories of plastic: thermoplastics. The only truly nonrecyclable plastics are of the other category, thermoset plastics, which include polyurethanes and epoxy resins; the polymers can't be pulled apart and rearranged once they've been "set." So these plastics, which are used in a broad range of products, from circuit breakers to motor components, tool handles, and the original plastic product, billiard balls, can't be recycled profitably because sufficient market demand does not exist to make it economically viable.

Here is the breakdown of the types of thermoplastics, according to the system that was crafted in 1988 by SPI for labeling the different types with numbers:

1. PET (Polyethylene terephthalate): used for soda and water bottles and lots of jars

2. HDPE (High-density polyethylene): used for milk jugs, detergent and shampoo bottles

3. PVC (Polyvinyl chloride): used in lots of household goods, and also added to many beauty products

4. LDPE (Low-density polyethylene): used in bubble wrap, shrink wrap, and bread bags

5. PP (Polypropylene): used in packaging, pipes for construction, and textiles

6. PS (Polystyrene): can be both rigid and made into a foam; in the rigid form, it's used for plastic cups, silverware, and plates, and in appliances, electronics, auto parts, gardening pots; in foam form it's used for clamshell food packaging, packing peanuts, and insulation

7. Other: how this category is generally and unhelpfully referred to; plastics in this category have been the most difficult to work out recycling methods for and include nylon, acrylic, and fiberglass, and films like cellophane

Investing in advanced plastics recycling technologies as fast as possible is one of the most important mandates for tackling the plague of plastic pollution. Additionally, promising progress is also being made in creating new types of biodegradable plastic. The commercialization of biodegradables is further off than the new recycling technologies, but many start-ups have proven the concept. Dutch company Avantium, for example, is seeking to commercialize a plastic made from sustainably grown crops that can be shaped into bottles, and Carlsberg, Coca-Cola, and Danone have signed on to the new technology. Molly Morse, a graduate student at Stanford, has developed a biodegradable plastic pellet that can be shaped into a variety of materials. Her initial plan was to reduce the plastic waste left over from the construction of disaster-relief housing, but she quickly realized her material break-through could also be used to replace plastic on a larger scale. Her company, Mango Materials, is producing this novel form of plastic through a process that uses methane, providing an added environmental bonus by removing a particularly noxious greenhouse gas from the atmosphere. Most important, and this stands true for all material types, not just plastics, we need to upgrade the definition of "recyclability" from "technically able to be sorted and recycled into a new product" to one that states that something is only fully recyclable if that material type is profitable for a municipality and related recycling facility to recycle.

Reducing, Refilling, Replacing

The public must keep pushing for bans or fees on any material types or products, plastic bags and foam, that are not profitable to recycle. Not doing so leaves the public responsible for the cost of disposal. The duplicitous efforts of some virgin plastics producers to thwart them are vigorous. Another tactic used of late is to work with state legislatures to pass bills banning the passage of bag bans, called preemption laws. Specially formed for that purpose is the Bag the Ban project, launched by the American Recyclable Plastic Bag Alliance and run by Matt Seaholm of the Plastics Industry Association. This is a perfect case of what reporter Tim Dickinson refers to as the organization creating "a nesting doll of front groups," in an article about the industry's lobbying in *Rolling Stone.* As of this writing, the tactic has assisted passage of preemption laws in Tennessee, Florida, Wisconsin, Indiana, Iowa, Michigan, Mississippi, Missouri, and Arizona.

In 2020, Closed Loop Partners convened a number of the world's largest retailers, including Walmart, Target, Kroger, and CVS Health, for a design challenge that will identify alternatives to the plastic bag. Today, people are often unaware that the plastic bag was not introduced until 1975. Rest assured, there was plenty of commerce prior to 1975, and as communities that have already banned plastic bags have demonstrated, any commerce that exists today will not be interrupted by the disappearance of plastic bags. What will disappear is the enormous tax burden communities face for landfilling plastic bags, the cost to local recycling facilities from the damage caused when plastic bags clog their machinery, and the eyesore when we see the litter of plastic bags at beaches and parks.

Another front in the fight to reduce the amount of plastic produced is creating planet-friendly plastic alternatives. One that is already taking off is mushroom foam packaging. The founders of New York State–based Ecovative Design figured out a way to grow mycelium around biological waste products, such as wood chips and corn stalks, which then can be hardened to make a foam packaging that's as sturdy as plastic foam, is entirely nontoxic and highly water resistant, can be grown into any shape, and is 100 percent biodegradable, so it's great for composting. They've made packaging for wine bottles by having the mycelium grown around a wine bottle that's then removed. British audio equipment maker Bowers & Wilkins commissioned packaging shaped tightly around its speakers, and both Dell and IKEA have ordered packaging from the firm. They're expanding their operations through a partnership with the Paradise Packaging Company in Paradise, California, as well as licensing rights to use their process to a number of firms overseas.

The plant-based plastics alternatives of cup makers SoluBlue and Footprint, who competed in the NextGen Cup Challenge, are only two of many already in production. Along with their cups, Footprint is also developing replacement trays for food producers and grocers to use for meats, fish, and produce. Footprint is in discussion with a number of top brands, including Bose, Philips, and Target, to create packaging for them. Work is under way pursuing many other possibilities, such as alternatives made from milk protein and wood lignin.

As for consumers, we can avail ourselves of a host of sustainable plastic product replacements that are packaged in sustainable plastic alternatives. Just about any kitchen item, personal care product, or other plastic throwaway is offered by a wide range of shops and online sellers, such a Life Without Plastic and EcoRoots—from bamboo toothbrushes with plant-fiber bristles, to deodorant packaged in cardboard cylinders with cardboard application rollers, a wide array of reusable produce bags and containers for groceries, and biodegradable dog and cat poop bags. As *The New York Times* reported in the article "Life Without Plastic Is Possible," a number of rigorous plastic abolishers write blogs and have published books with tips about the options, such as *Plastic-Free: How I Kicked the Plastic Habit and How You Can Too* by Beth Terry, who also offers a wealth of advice and product information on her website, My PlasticFreeLife.com. Getting to actual zero plastic waste is probably not possible. Beth Terry, for example, has found that she can't get pharmacists to sell her medications in refillables she would bring in. But as more and more nonplastic options make their way to market, we can vote ever more powerfully with our wallets for plastic eradication.

The major consumer goods companies have recognized the market demand, making significant efforts to eliminate any low-value plastics from their product lines. Unilever designed a cardboard box something like a beer six-pack box for its Solero ice-cream pops, so they don't need to be individually wrapped in plastic. Speaking of beer, both Carlsberg and Guinness have announced they'll no longer use plastic-ring six-pack holders. Nestlé started packaging its Nesquik drink powders in paper rather than plastic, and its Institute of Packaging Sciences is exploring possibilities for plastic replacements.

A number of leading brands are also testing models for another of the main approaches to reducing plastic production: refillables, which Chilean-based Algramo has been achieving such success with. Coca-Cola invested $25 million in designing a standard PET bottle for all of its sodas and $400 million to create the infrastructure in Brazil for a returnable bottle deposit system, which reportedly achieved a 90 percent return rate. The bottles can be refilled twenty-five times, after which they are recycled.

An innovative alternative model for refillables has been implemented by the Loop program, a project spearheaded by recycling company TerraCycle in partnership with several corporations, including Coca-Cola, Nestlé, PespiCo, and Unilever. Loop allows customers to order popular products from participating brands online, such as Häagen-Dazs ice cream, Tide detergent, and Seventh Generation cleaning products, all of which come in specially designed refillable packaging. The customer is charged a deposit on the packaging, which is returned when they send it back or kept on deposit if they order the package to be refilled. Their items arrive in a specially designed container like a large, soft cooler, and return is by a scheduled pickup.

An Infinity of Recyclability

In stark contrast to the mass-producers of plastic, the first mass-producers of glass, the ancient Romans, fully appreciated how precious glass was, and that it should be recycled. In fact, glass can be recycled infinitely. The high value the Romans placed on glass can be seen in the ruins of an ancient ship, the *Julia Felix*, discovered in the Adriatic Sea in 1986 and dated to around 300 CE. The ship's cargo was exceptionally well-preserved—clay vases containing olive oil, wine, sauces, and assorted fish products. Attesting to the extraordinary ingenuity of the Romans, the remains of a pipe and a pump system were also discovered, which siphoned seawater up into a glass aquarium on the ship that held up to 440 pounds of live fish.

Also found in the *Julia Felix*'s hold was a large barrel of glass shards for recycling, and such crates of broken glass have been found near remains of kilns in locations all around the vast expanse of the Roman Empire. Cases full of glass ingots—little square chunks—have also been found, which were made by the Romans expressly for recycling. They could be safely transported over bumpy roads and tossing seas and then melted down for making into new products.

Which speaks to one of the problems with replacing plastic bottles and containers with glass ones. Even as durable as glass is, unless it's quite thick, it easily breaks. That means containers made for transport—of water, soda, wine, or what have you—must be pretty thick, making them much heavier than plastic versions. The typical glass bottle, for example, is seven times heavier than its equivalent in plastic. The weight of glass makes its transportation a good deal costlier, and also increases its carbon footprint. The centralization of the bottling business in the post–World War II years, which was formerly so locally distributed, also limits the practicality of glass bottles as returnables. But on the flip side, glass containers can be reused at least fifty times without degrading.

There is at least one product, though, for which glass returnables may make particularly good sense: craft beer. Beer bottles are still common for beer because glass is so preservative. It doesn't allow oxygen to permeate, making it a great material for containing fermented liquids like beer and wine, which exposure to oxygen quickly turns bad. So preservative is glass that a wine bottled in 300 CE, discovered in 1867 in a Roman aristocrat's tomb in Speyer, Germany, has retained its contents well enough that "microbiologically it is probably not spoiled," according to wine expert and professor Monika Christmann. Though, she's quick to say, "it would probably not bring joy to the palate."

With the resurgence of local brewing, the Oregon Beverage Recycling Cooperative (OBRC), got the idea of instituting a returnable beer bottle system, its "BottleDrop" program. Because Oregon passed a bottle deposit bill way back in 1971, the state has a well-developed infrastructure for bottle collection. I was fascinated to learn from OBRC's head of market development, Joel Schoening, that the progam's development was funded by Coca-Cola, PepsiCo, Nestlé, and Columbia Distributing in Washington State. Due to the bottle bill, they have an interest in getting returns, a great testament to the importance of passing more laws.

After much deliberation with brewers, who all had their preferences for shape and color, the OBRC designed a brown bottle designed to be refilled, and the program is recouping 81 percent of them, with twelve breweries participating. The bottle is now also being used by two wine producers in the Willamette Valley. At 500 milliliters, the bottles are smaller than the standard for wine, more like the size of the classic French bistro wine bottle. Joel Gunderson, the manager of the Coopers Hall Winery, which also runs a restaurant, told me many restaurant customers say it's perfect for sharing over a meal. I asked if patrons found them odd, and he said, no, Oregonians are what Gunderson calls "moral shoppers," who see the bottle program as "salmon safe," an Oregonian shorthand for environmentally friendly. Happily, so are lots of consumers in other hubs of microbrewing. A sign of the potential for scaling of the bottle program is that all of the state's fifty-six Safeway stores are now selling two of Coopers Hall's wines in the bottles.

Beer comes in bottles of different colors because each has advantages for different beers. Amber glass is the best for preventing beer getting "skunked" due to exposure to sunlight. Clear glass shows off the golden purity of pilsner. The problem is that the different colors must be separated for recycling, otherwise the new glass will be murky. But in the U.S. all colors of glass are mixed together in collection, and there hasn't been an economical way to separate them. The result is that less than a third of bottles and jars sold in the U.S. are recycled. But a simple change in the system for collection would fix the problem—providing bins with separate compartments for each color. That's what they do in the Netherlands, which has a glass recycling rate over 90 percent.

Consumers there have fully accepted that they've got to sort glass by color. When I went on a tour of Amsterdam with a Dutch official involved with the city's circular economy programs, to research their system as a possibility for New York, I asked him what happens when someone puts the wrong color glass in a bin. Do they get a big fine? "Well," he replied, "No one would ever do that." I said, "But sometimes they must, and then what do you do? How do you sort them out?" He answered, "We don't. No one ever does that." They don't do it in Germany either, which has a glass recycling rate of 98 percent.

Because the economic challenges the complexity of color sorting creates for glass recycling, another alternative to plastic is aluminum. It is by far the most valuable commodity in the recycling industry, is infinitely recyclable, and already boasts the highest recycling rate of any commodity used in consumer products. As always in advancing circularity, we should move on multiple fronts. The key is to ensure that every product or package has a clear value in the recycling stream before it ever hits the store shelf.

Cleaning Our Waterways

While some ingenious methods have been developed to remove plastics from our oceans, rivers, and lakes, it is critical that we implement systems to ensure that plastics never get into our waterways in the first place. As soon as a piece of plastic enters our waterways the damage to our ecosystem and our health begins. Cleaning it out simply minimizes the damage, and most plastics recovered from our waterways have already degraded to a point where they can't be recycled into a new product. In 2016, the Ellen MacArthur Foundation published a report, "The New Plastics Economy," in which it issued a prediction that resounded around the world: "If the current trend" of plastic waste continues, "there could be more plastic than fish (by weight) in the ocean by 2050." With all the brilliant efforts we've surveyed here to combat the plastic waste scourge, we may very well be able to clean up our coastlines and rivers, obliterate the massive plastic gyres, replenish fish stocks, and restore the health and beauty of the marine ecosystems that are so vital to life on the planet.

Gold Mines in
Our Hands

O N APRIL 11, 2019, I was watching live coverage of the Kennedy
Space Center in Florida as the twenty-seven Merlin engines
of the SpaceX Falcon Heavy megarocket were fired up. Pure
white steam billowed out of the rocket's core, and at 6:35 p.m., the
engines ignited with a thunderous boom as 5.1 million pounds of
liquid oxygen propelled the world's most powerful rocket skyward.
Accelerating from zero to 23,663 miles per hour in two minutes and
thirty-nine seconds, the rocket streaked across the Kármán line, the
border between Earth's atmosphere and space. A moment later, in
an unprecedented feat of spacecraft engineering, the Falcon Heavy's
two side booster rockets separated from its third core rocket, turned
around, and began to descend.

As they landed side by side in perfect symmetry, the crowd
erupted in cheers. One minute and fifty-seven seconds later, the
center core rocket, having released its payload into low Earth orbit
and performed its own masterful flip, touched down on the ship
Of Course I Still Love You, anchored a mile off the Space Coast in
a roiling Atlantic Ocean.

The trifecta of successful return journeys marked a break-through in bringing circularity to electronics—making one of the most intricate and costly machines ever built reusable. By re-furbishing and relaunching rockets, SpaceX founder Elon Musk and his team have brought the cost of launches down to less than half that of building a new rocket, beating the pants off longtime space engineering leaders like Boeing and Lockheed Martin.

Musk's strategy of reusability paid off handsomely when SpaceX secured a major deal with the U.S. Defense Department for the second Falcon Heavy launch, in late June 2019, to deliver twenty-four satellites into orbit. Also included in the payload was a solar-powered spacecraft created by the nonprofit The Planetary Society. The tiny craft—the size of a loaf of bread—is the first test of a grand vision to use only the sun's clean power to fuel the exploration of space. Musk, who has revolutionized travel on Earth with his fleet of electric Teslas, is now helping sustainably power transit to our furthest frontier.

Cost saving isn't the only advantage of reusability. Reuse as-sures quality. The company can inspect every iota of a returned rocket to see where it might have any near-miss defects, made clear by its space journey, and then correct for them. In fact, repeated flights impress insurers. As NASA administrator Jim Bridenstine says, "When you use a rocket a second, third, fourth or fifth time, insurance rates actually go down, not up." All of which is why Elon Musk calls reusability the "Holy Grail."

That Musk's engineering team could design a pathbreaking modular engine system that makes engines easy to repurpose and improve a heat-shield recipe of NASA's to make shields that with-stand higher heat and cost considerably less—shows how much can be done to make even the most complex devices durable and reusable. If they can figure out how to make the world's most ad-vanced rocket so economically and restore it so quickly for reuse, why does repairing an iPhone or laptop or flat-screen TV still seem to be a complex and cumbersome endeavor?

The price of repair is so high, and the types of repair that can be done are so limited, that even with the soaring cost, buying a new phone usually seems the obviously better option. And it's often the only option.

The sad fact is that our consumer electronics aren't built for repair, and the result is ever-growing mountains of electronic waste—e-waste—the fastest growing flow of castoffs to landfills worldwide. Every year, an estimated 50 million tons of electronic devices are thrown away globally—the equivalent of all the planes that have ever been made for commercial flight—representing mountain upon mountain of lost value. In fact, the value of electronic waste is estimated as $62.5 billion. Yet only 15 percent of discarded electronics are sent for recycling annually in the U.S. The figures are better for Europe—ranging from about 40 to 50 percent in most countries—and in Asia it is over 60 percent in most countries. That's because their governments require sellers and manufacturers to take charge of recycling specified percentages. Even so, only an estimated 29 percent of e-waste globally is recycled annually. The waste of increasingly rare precious metals is astounding.

Consider the substantial amount of gold in mobile phones. Federico Magalini, a leading expert on the electronic waste problem, points out that the ratio of gold to be extracted from cell phones per ton is a whopping *eighty times* higher than from mining gold ore. One ton of phones can yield about twelve ounces, and extracting the gold from one million phones produces seventy *pounds* of gold (sixteen ounces per pound). Meanwhile, gold mines are going to extremes to reach an ever more elusive supply.

The Mponeng Gold Mine in South Africa is often called the world's deepest man-made hole. It's much more than that. Depending on your perspective, the mine is either an engineering wonder or a humanitarian and ecological travesty. To get to its 236 miles of tunnels, 2.5 miles belowground, miners must take three elevators. The journey takes an hour. With the temperature in the shafts reaching 140 degrees Fahrenheit, to keep the miners alive, icy water is pumped throughout the tunnels and blown over them by enormous fans, bringing the temperature to a still-steamy 85 degrees. The mine uses six thousand tons of ice per day.

All of this is to reach the spindly tendrils of a vein of gold ore that is a mere thirty inches thick at its widest; within that ore, the gold is diffused in tiny flecks rather than a solid mass. It takes a ton of ore mined to produce .35 ounces of gold, and the mine extracts an average of 250,000 ounces of gold a year. That's over 700,000 tons of rock pulled up. At a cost of $779 an ounce of gold extracted, with gold selling for a fluctuating price hovering around $1,300 an ounce, the mine still makes an impressive profit. But the question is, If the world is so hungry for gold that creating an infernal underworld and mining at such great expense makes economic (if not ecological and humanitarian) sense, then why do we allow so many of our cell phones, tablets, laptops, and TVs to end up in landfills? Phones, tablets, and laptops also contain appreciable quantities of other precious metals, including copper, silver, and palladium. We should be mining all those devices!

One company, Electronics Recyclers International (ERI) is leading the way in showing how lucrative recovering all the lost value of e-waste is.

The Gold Mine of E-Waste

John Shegerian's entrepreneurial career is an illustration of the wisdom Steve Jobs famously shared in a commencement address at Stanford University—that we can't connect the dots of our lives going forward, only looking back on them. John and his wife, Tammy, are the founders of ERI, one of the largest and most technologically advanced e-waste recycling businesses in the United States. I got to know John when I was building Recyclebank, and was invited to a meeting of recycling experts convened by aluminum giant Alcoa. John was there because ERI has a partnership with Alcoa to provide it with recycled aluminum, just one of many brilliant business moves the Shegerians have made.

His burning ambition as a child was to become a jockey, but at a young age it was already apparent he would be much too tall for that. With a belief in somehow finding a way that has characterized each of his business endeavors, he decided instead he'd become a harness driver, taking a job mucking stalls at a track north of New York City, where he grew up, as a first step. He became the youngest professional racer in the U.S. at age sixteen, and set a world record in the sport when he was seventeen.

His father decided to get in on the racing business too, purchasing the stallion Noble Darcy. Setting the pace for John in seizing serendipitous opportunities, when his father sold Noble Darcy as a stud horse to a Belgian manufacturer of windmills, he decided to make his own big leap: bringing wind power to the U.S. This was in 1977, and the company he founded, WindMaster, thrived. That is, until Ronald Reagan became president and did away with tax credits that had supported investment in the technology. The company folded. Rather than taking a negative lesson from that experience, John was struck by the possibilities of entrepreneurship to solve problems.

Before founding ERI, he cofounded Homeboy Industries with Father Gregory Boyle, which employs formerly incarcerated individuals. They started with Homeboy Tortillas, a small tortilla stand in Los Angeles's Grand Central Market, which was so successful it was featured as the Christmas story of the year by CNN. After marrying Tammy, he took what he'd learned about the food business and decided to open a restaurant and brewery in her home city of Fresno—where he produced Bulldog Root Beer and donated the proceeds to Fresno State's Craig School of Business.

Through involvement with the school, he learned of the steep costs of college and founded FinancialAid.com, which thrived. From the head of the FinancialAid.com sales group, he heard that her best friend was running a struggling company, Computer Recyclers of America; she thought it could use John's help. John and Tammy were so impressed by the potential of e-waste recycling that in 2005 they decided to sell FinancialAid.com and take over the recycling business, moving it into a former rag recycling facility in Fresno that had been vacant for twenty-five years. In their first month of business, they recycled ten thousand pounds of e-waste. That was in April 2005. In April 2019, they recycled 30 million pounds at eight facilities around the country.

John and Tammy have continuously spearheaded major innovations, in both the mechanics of e-waste recycling and in their business model. In partnership with the Bloomberg administration in 2013, they launched the first collection system in the world for e-waste at residential buildings. ERI now picks up from eight thousand buildings in New York City, allowing 3.2 million residents to safely dispose of their devices simply by scheduling collection.

John and Tammy didn't start ERI with any fancy equipment. At first, employees dismantled devices by hand. But soon they decided to look for a machine to speed the process by shredding electronic devices. Tammy took charge of traveling to Asia to look for good shredding machines, because Asia is ahead of the curve in e-waste recycling, but none were available. So she suggested to John that they hire an engineer to create one. They did, and watching the machine munch up everything from the huge flat-screen TVs to industrial-scale computer processors and all manner of smaller devices is a marvel to behold. They make their way by conveyor belt into its massive maw, coming out as finely granulated bits that are sorted by material.

John and Tammy also developed an innovative approach to extracting the gold, silver, and other precious metals from circuit boards. Soon, members of the founding family of Korea's electronics giant LG learned of the process and came calling. In addition to LG, they run a copper-smelting business, and they have invested in the company and partnered with ERI to provide extracted circuit boards and the technology for mining them—because, as they told John, mining precious metals from electronics is a much more reliable, socially responsible, and economical means of obtaining them than contracting with mining firms. Indeed, a study conducted in China found that mining copper, gold, and aluminum from ore costs thirteen times more than recovery from mining of e-waste.

The latest pathbreaking move ERI has made is in robotic sorting. In 2018, they became the first e-waste recycling facility to begin utilizing robots.

Recycling Robots to the Rescue

When you meet Matanya, there is no doubt you are speaking with someone who is thinking in a different dimension. As a graduate fellow at the California Institute of Technology, one of the world's leading robotics research centers, he pursued a childhood dream. Watching *The Jetsons* and *Transformers* as a kid, he was thrilled by the prospect of helping build personal robot companions, and while at CalTech he took part in some of the most sophisticated competitions in that arena. For the ARM-S project, sponsored by the U.S. government's Defense Advanced Research Projects Agency, he worked on building robots that could open a bag with a zipper and change a car tire; for another project, he helped assess how well robots would have been able to conduct cleanup after the Fukushima nuclear reactor disaster.

That work has quite a ways to go yet. As he explained to me, so many tasks that humans can do effortlessly still pose insurmountable challenges for robots. Even walking: "Robots stub their toes all the time," he says. Matanya brilliantly turned his focus to robots for recycling, because as a specialist in machine vision, he perceived that a confluence of technologies had emerged for robots that can see well enough to sort recyclables, and he knew they'd be able to do so a good deal faster than even the most experienced human sorters.

By combining his robots with AI, he's enabled those robots to sort over fifty items a minute. Before long, he expects they will be able to sort green from brown and clear glass, and deftly pick bottles up, facilitating reuse. Mantanya's a big fan of circular principles, and his business model was premised on making the robots easy for customers to repair, with AMP technicians guiding on-site work through remote assessments.

From Hype to Planet Healing

We've been promised that robots with humanoid bodies are just over the horizon going back for decades, along with all sorts of other artificial intelligence, big-data analytics, nanotech, and quantum-computing marvels. Back in 1995, Nicholas Negroponte, then the head of the MIT Media Lab, wrote in his bestselling book *Being Digital*, "Early in the next millennium, your right and left cuff links or earrings may communicate with one another by low-orbiting satellites." What they'd be saying to one another, it's hard to figure. While that hasn't come true, when it comes to low-orbiting satellites he was bang on. Elon Musk's rockets are launching them fast and furiously. Amazon plans to create a network of 3,236 of them, to bring internet service to all the regions of the world still deprived of it.

As for robots, Negroponte thought we'd have "digital domestics with legs to climb stairs and hands to carry drinks," and though such robots are still to come to our homes, delivery bots have arrived. Starwood Hotels has deployed robot butlers it calls Botlrs, which deliver room-service meals, towels, and toothbrushes to guests. They were so popular during the trial pilot that guests were wracking their brains to think of more things to have delivered. Boston Dynamics has created a robot, the Atlas, that can do backflips. So, perhaps robots will be flipping themselves up the stairs rather than walking.

Predictions about technological innovation are notoriously tricky. Experts on the new wave of tech sharply disagree on some points; for example, some predict a "robot apocalypse" that will wipe out whole professions, while others argue artificial intelligence–enabled devices will enhance our work lives. What's not in dispute is that the confluence of developments is opening up a wild frontier of opportunity for inventors and entrepreneurs to play a transformative part in creating the much-touted Internet of Things (IoT). As Dr. Kai-Fu Lee, one of the world's leading authorities on AI, wrote in his bracing book *AI Superpowers,* "We stand at the precipice of a new era, one in which machines will radically empower and/or violently displace human beings," stressing that "it's now time for entrepreneurs to roll up their sleeves and get down to the dirty work of turning algorithms into sustainable businesses."

He didn't mean sustainable as in green, rather as in viable— but green applications are flourishing. Many innovators are using the new "smart tech" to further the cause of circularity and combat the ravages of climate change. The cost of a vast range of sensors that can remotely monitor the functioning of devices is coming down so rapidly, along with their size—with some now about the size of a sesame seed—that they'll be embedded in more and more products to spot needed repairs and perform "predictive maintenance," as Michelin has been doing with tires and Rolls-Royce has done with its Power-by-the-Hour program. My favorite story of repair monitoring is from a company called SweetSense, which sells devices at low cost that monitor water-well pump equipment in the developing world, with the data communicated by satellite. By sounding an alert about problems, the devices have brought the failure rate of pumps down significantly, which in a future of increasing droughts will be a vital lifesaver.

Sensors, combined with technology for tracking products throughout their life cycle (primarily radio frequency identification [RFID] tags and block chain), are opening up great possibilities for the product-as-a-service model, also called "servitization." One major electronics player exploring the model is Samsung, beginning with leasing TVs. Matanya Horowitz, always innovating, is now leasing machines as well as selling them, making them affordable to smaller operations so they can greatly ramp up their volume of recycling.

Connectivity and monitoring enabled by the technology is

allowing not only for repair but for optimization of performance. The Spanish passenger railroad service Renfe has used IoT technologies to do predictive maintenance on trains and also monitor on-time performance using the data collected to improve service so dramatically that the firm reportedly achieved a record of just one significant delay in more than twenty-three hundred trips. Due to that improved performance, the service has earned a 60 percent market share for travel between Madrid and Barcelona, which had previously been dominated by airlines. A number of technology leaders, such as Siemens, are offering packages of IoT tools for manufactures to snap up, making it easier for them to apply the technologies. The possibilities for making all manner of electronics products—from washing machines to refrigerators, air conditioners, and small appliances—more energy efficient and longer lasting are game changing.

Inventive combinations of the technologies are also helping to save the planet's biodiversity and protect its forests. One acclaimed initiative is Eyes on the Forest, which makes use of relatively inexpensive off-the-shelf drones combined with the massive data capabilities of Google's Cloud service and its mapping technology, as well as free satellite imagery provided by NASA, to monitor illegal logging in the rain forests of the Indonesian island of Sumatra. Dubbed the "Emerald of the Equator" due to its lush forests, which once covered the island from shore to shore, Sumatra has lost over two thirds of its lowland forests in the last twenty-five years, much of it to provide palm oil but also for paper pulp. That's been devastating not only when it comes to carbon capture but also for the island's majestic species of tigers, elephants, and orangutans, which are all now critically endangered.

Eyes on the Forest was founded by the World Wildlife Fund in partnership with two local NGOs to collect evidence of illegal operations. A team of undercover investigators fans out into the forest to take drone footage of clear cutting, documentation of which has been a dangerous business—one investigator was abducted, beaten, and held hostage by a group of loggers for several hours (but kept right on with the work after that experience). The drone surveillance has been critical to convictions against companies and corrupt government officials looking the other way about logging, and forced one of the largest global paper companies, Asia Pulp & Paper, to publicly pledge to stop its illegal logging. Google provided a grant to Eyes on the Forest for a developer to use Google Earth mapping technology, combined with the NASA satellite data to create a website featuring photos and a detailed interactive map of the forest, in order to monitor cutting and raise public awareness. It is available for viewing at https://maps.eyes ontheforest.or.id.

Another project making use of a combination of off-the-shelf technology, such as GPS monitors and solar-powered sensors and government-developed technology is creating the "Internet of Animals." Called ICARUS, International Cooperation for Animal Research Using Space, the project is attaching small, solar-powered sensors to large populations of wildlife, from insects to birds, lions, tigers, and bears, and will monitor their migration patterns and habitat degradation via equipment installed on the International Space Station. This will provide much more refined information to conservationists. Those of us in the broad public will be able to log on to an app on our phones and check in on specific birds, dolphins, or sea turtles we're interested in tracking, perhaps one we've rescued from a death by plastic waste.

Transformative innovations in solar power, in particular, are rapidly emerging. One is the creation of two-sided solar panels that use sensors and tiny engines to move through the course of the day and optimize the angle of light hitting them, like solar sunflowers. One model being produced by Solar Energy Research Institute of Singapore showed in tests that it would generate 35 percent more power, while at the same time reducing the cost of producing it by 16 percent.

Even more potentially transformative: the achievement of Heliogen, a start-up backed by Bill Gates, among many others, and founded by Bill Gross, who is well-known as the brainiac behind

Idealab—one of the premier technology innovation incubators in the world. The engineers of Heliogen have figured out how to use AI to configure a set of mirrors so they direct sunlight into such a concentrated beam that it can generate heat of 1,000 degrees Celsius, which is about a quarter of the heat of the sun's surface. It's hot enough to be used in the manufacture of steel, cement, and glass, which, as we'll see in the next chapter, would in just that one stroke substantially reduce carbon emissions.

One last hopeful development to mention here is a means of recycling solar panels, which is now economically prohibitive at large scale, leading to mounds of them beginning to make their way into landfills. A Belgian research study conducted at the University of Liège is working on a method for extracting silicon from panels, which could then be used in electric-vehicle batteries, while also purifying the glass in panels for reuse.

Obsolescence Pays

So many of the devices thrown away are working just fine, and in many cases could keep working well for many more years. Why aren't we seeing the true value of our electronic wonders? One answer is planned obsolescence.

Think of the Apple "Batterygate" scandal. The company was forced to admit it was intentionally slowing down the operation of some older iPhone models in 2018, shortly in advance of the launch of its newest models. Some technology experts mince no words about Apple's obsolescence intentions, such as *TechCrunch* reporter Seth Porges, who calls the iPhone "a slam dunk of planned obsolescence." But the planning of the obsolescence of tech devices has been devilishly hard to prove.

Many electronic companies calculated that selling more new products trumps selling a mix of new and used ones. Yet the industry that ordained obsolescence the new high priest of business has proven the business of restoring and reselling can be a great boon: the car industry has come full circle. It's building cars to last again, and has thrived by creating the robust certified pre-owned business. The preowned market has become so strong in fact that *Automative News* calls it a "wheel of fortune that keeps profits on the upswing" even though purchases of new cars are down. The average age of cars on U.S. roads is now 11.4 years. Compare that to the goal GM once set. In 1955, the company's head of design happily reported that the average length of owner-ship of a new car in the U.S. had dropped to just two years, adding, "When it is one year, we will have a perfect score."

Building for Longevity

One impressive achievement is that of Dutch start-up Fairphone. The company offers a modular smartphone built for longevity, aiming for a minimum of five years. It's elegantly designed and can easily be repaired by swapping out components, which owners just snap into place. The "fair" in the name refers to the company's mission to use only fair-trade precious metals in its phones: Founder Bas van Abel was inspired to create the phone by his own experience of the travesty of the mining of gold and other precious metals. "I saw guys digging sixty-meter holes into the ground and staying down there for days without any protection just to get this mineral out of the ground," recalls van Abel in a TED talk. He's referring to coltan. It's one of the "conflict minerals," so called because the trade in them has funded armed militias that have wreaked havoc in the Congo and elsewhere in Africa. The conditions in these mines are appalling, some toxically festering pits in which children pan for gold. Crude mine shafts are often so poorly reinforced that collapses regularly occur.

Van Abel traveled to the Congo in his role as creative director of the Waag Society, which wanted to bring public attention to the conflict-mineral problem. Though van Abel had never designed a phone, and hadn't even owned a smartphone, he thought making a phone that used only fair-trade metals would be the best way to raise awareness. About his lack of expertise, he told a reporter, "naivety is a great catalyst for discovery."

Van Abel crafted a business model for the Fairphone that was impressive enough to garner funding from the Dutch phone manufacturer KPN, as well as a prestigious spot in the incubator program run by London investment firm Bethnal Green Ventures. An outside investor also contributed $400,000. A website featuring a phone prototype was launched in 2013, with a public advocacy campaign offering the phone for sale, and within three weeks, van Abel recalls, "we had three weeks when we sold over 10,000 phones, we had 3.5 million euros in the bank account, and we didn't know how to make phones." Within three months, orders were up to 7.5 million euros. By a year and a half, he'd hired forty people and sales totaled sixty thousand phones, making the Fairphone the highest-earning crowdfunding campaign in Europe to that date and his operation the fastest-growing European start-up.

The company released its third model in early 2020, with replaceable battery, display, camera, speaker, and top and bottom circuitry modules for order on its website. It also has a headphone jack, so no pricey earbuds are needed, and it comes with no charging cable because it's made to be charged with any USB cable. Coming out with a new model hasn't meant Fairphone no longer supports the prior model. In June 2020, the company announced that the Fairphone 2, which was released in 2015, could be upgraded with the Android 9 Pie operating system, with a two-year software maintenance warranty.

The achievement of the Fairphone team is all the more impressive in light of the failure of a number of other highly anticipated modular devices ever to hit the market. A particular disappointment was the discontinuation in 2016 of Google's Project Ara, which was developing a phone that was to include modules not only for cameras and speakers, but specialty features like a receipt printer for purchases, a laser pointer, and even medical devices.

Modularity may need to start small, as former Motorola engineer and serial entrepreneur Baback Elmieh believes. He founded Nascent Objects, which offered modular components for building one's own smart electronic devices. Facebook bought the company in 2016, and also hired Elmieh. He thinks modularity at large scale is coming, but that "it's going to be a small niche at the beginning." Just so, it's often small start-ups who build specialty markets that prove the concept of innovations for larger firms.

The Right to Repair Cannot Be Denied

One day in early 2020, Kyle Wiens, the cofounder and CEO of iFixit, was listening to coverage of the rapidly spreading coronavirus pandemic. He was galvanized when he heard that in Italy, ventilator machines were under such stress from higher use that they were beginning to break down. He knew iFixit had to snap into action. The company provides repair information, and "tear-down" videos demonstrating repairs online, for free, and sells spare parts to consumers for thousands of electronic devices. Wiens is one of the leading figures in the right-to-repair movement, which is pushing for legislation in the U.S. and around the world that would require electronics manufacturers to facilitate repair by consumers and independent technicians. He consulted with Bas van Abel about how to make the Fairphone optimally repairable, and he's been a thorn in Apple's side by making iPhone repair information easily downloadable and harvesting parts for sale. Wiens knew that the manufacturers of biomedical equipment generally restrict access to their manuals for repairing devices, charging high fees for training sessions they give to on-site hospital technicians called biomeds.

Manufacturers argue that only their technicians have the skills to work on their machines, forcing hospitals to wait for an authorized technician sent by the manufacturers, sometimes for days. But biomeds are highly trained, and in 2018 the Food and Drug Administration declared that they conduct "high quality, safe, and effective servicing of medical devices." Biomed Nader Hammoud of San Francisco's John Muir Health has spoken out forcefully about the problem. When I talked with him, he recounted a struggle he had to get a manufacturer to send him an eighty-dollar part. They were telling him that one of their own technicians would have to come to install it, which would've taken three to five days. Price tag? Four thousand dollars. But cost is the secondary problem. Devices can break down 24/7, and Hammoud has often had to rush to the hospital to fix a machine in the middle of the night. "It's not like you want to change the oil in your car," he says. "There is a patient attached to that device." Meanwhile, due to lockdowns, outside technicians couldn't travel to hospitals.

Yet as the pandemic raged, some manufacturers of ventilators continued to withhold repair guides. In response to a letter signed by 326 biomedical professionals and sent to two dozen manufacturers, Medtronic, GE, and Fisher & Paykel shared their repair manuals. But others held firm. Because, as University of Pittsburgh law professor Michael Madison, a specialist on the right-to-repair debate, explains, the profits from selling the machines are marginal and "the profit is in servicing." Fortunately, Kyle Wiens was on the case. He immediately launched Project BioMed, putting out a call worldwide to biomeds to share repair manuals. The effort focused at first on ventilators, anesthesia machines, and respiratory analyzers, but branched out to a huge range of devices, collecting over thirteen thousand manuals.

I asked Nathan Proctor of U.S. PIRG, a coalition of state Public Interest Research Groups that spearheaded the letter campaign, why Congress and state legislatures haven't acted to address this glaring problem. One issue, he says, is that lawmakers don't understand the technology well enough. I recalled the 2018 hearings in which senators "grilled" Mark Zuckerberg with such hilariously out-of-touch questions as, "A magazine I recently opened came with a floppy disk offering me thirty free hours of something called America Online. Is that the same as Facebook?" Nathan Proctor reminded me that so many products we might think require specialty repair skills can actually be remarkably easily fixed. He laughs that when he took the U.S. PIRG job, he started getting a flood of queries about how to fix items, but wasn't himself a tinkerer. Soon, he decided to become one. His favorite fix? A washer-dryer he'd bought that broke after thirteen months, of course under a twelve-month warranty. He had to build a rig to lift the machine up above his head—but he was then able to get it running by replacing a ninety-dollar part.

Legions of fix-it hobbyists aren't waiting for regulation. They're flocking to repair cafés all around the globe, either to work on their devices themselves or to get specialists to help them, all for free. The leader here is a nonprofit foundation, Repair Café, started by Martine Postma of Amsterdam, who hosted the first event in 2007. As of this writing, the organization's website listed 2,085 cafés along with a free manual about how to start one yourself. The goal is to inspire a joy of fixing things and to raise awareness of the throwaway culture's devastating consequences.

In Europe, the push for the right to repair has made great progress. The European Union Parliament passed a resolution on a longer lifetime for products in 2017, which urged all member states to "take measures to ensure consumers can enjoy durable, high-quality products that can be repaired and upgraded." In 2020, the EU went further, introducing laws that will require manufacturers to make replacement parts available for ten years after product launch and to make them easy to replace with simple, common household tools. Making repair manuals available is also mandated.

In the U.S., the movement scored a rousing victory when Massachusetts passed the first national right-to-repair law, supported by 86 percent of voters. The law led shortly thereafter to the major car manufacturers agreeing to allow independent mechanics to repair their cars. Right-to-repair bills have now been introduced in twenty states, and the Federal Trade Commission hosted a workshop titled "Nixing the Fix" in July 2019 to explore the nature of repair restrictions manufacturers impose and whether they violate consumer rights.

Predictably, industry lobbying groups, disingenuously named as ever, have been vigorously opposing all measures. One is the Consumer Technology Association, which is actually an organization of technology companies. The Security Innovation Center, which argues for what it calls "sound repair policies," lobbies against the bills by asserting that the right to repair would jeopardize consumers' security. Yet, as *Consumer Reports* spotlighted, when the Washington State legislature asked for examples to back up that contention during hearings about the bill proposed there, state representative Jeff Morris says, "None were ever provided."

I spoke with technology security expert Paul F. Roberts, publisher and editor in chief of the Security Ledger website and founder of the organization Securepairs, a group of leading security professionals who advocate the right to repair. The objection to allowing repairs isn't based on concern about security, Roberts says, but on concern about profits. "The divide is between large corporations," he tells me, "and their customers. Right to repair is bipartisan."

Thankfully, some leading brands are making a good business case for facilitating repair. Samsung has a large network of authorized independent repair technicians and is working with them and other suppliers to harvest used phones for refurbishment and resale by the company for its certified preowned business. Going a step further, Samsung has also explored potential for upcycling phones. In a companywide competition, product engineers were asked to propose ways phones could be upcycled. One idea selected for prototyping was a device for conducting eye exams, which takes the place of machines that generally cost on the order of three to four thousand dollars. In the developing world, that price tag has prohibited purchase by many eye doctors. In a pilot program in 2019, Samsung provided ninety of the machines to doctors in Vietnam, who successfully performed exams with them on fourteen thousand patients. That's the most heartening upcycling story I've ever heard.

Another leader is Hewlett Packard Enterprise (HPE), which

was split off from Hewlett-Packard in 2015. It operates two enormous Technology Renewal Centers, one in Massachusetts and the other in Scotland, at which it refurbishes not only HP computer equipment but that of other brands as well. Lots of it. The Scottish facility, for example, reports that it processes about one hundred thousand IT assets a month. The manager of HPE's Scottish facility, Jackie Rafferty, says, "Certain assets here are like classic cars; they can become very valuable in the future." NASA can attest to that. When NASA needed to find a replacement for a sixteen-year-old computer critical to running some of its systems, they came to Rafferty's team. They found one stored away "with the packaging still on it." If a sixteen-year-old machine is keeping NASA programs running, just imagine all the valuable work our much more powerful newer machines could be doing in their next life.

9

Building to Heal

S EVENTY-FIVE MILES OFF THE coast of San Simeon, California, where William Randolph Hearst built his garishly sprawling 115-room, 68,500-square-foot Hearst Castle atop La Cuesta Encantada, and 5,000 feet below the Pacific's placid surface, an altogether more awe-inspiring structure than Hearst's mansion has been discovered. The Davidson Seamount is a massive dormant volcano, 26 miles long, 8 miles high, and towering 7,480 feet, sometimes called an underwater megacity. Monterey Bay National Marine Sanctuary biologist Chad King, who has explored the mount by remotely operated vehicle, describes the experience as "like hanging from a helicopter at night with a flashlight attempting to describe Manhattan. You can get some idea of the communities and life there, but you see only a small part of what is happening."

The mount is home to a teeming ecosystem of creatures, from a colony of thousands of shimmering lavender-colored octopuses, dubbed "octopaloosa" by researchers, to old-growth forests of fluorescent green and incandescent alabaster corals, and colonies of spherical, translucent jellyfish that look like futuristic spaceships. Not only does the mount support at least 168 species, with likely hundreds more yet to be observed, it also sends streams of nutrients up toward the surface and creates ocean current patterns that assist the flourishing of flora and fauna that hug the coastline.

When I watched video of a robotic submersible exploring the murky depths of the mount, it caused me to wonder: What if we humans had built our own megacities with something even approximating the ecological brilliance nature brings to her creations? What if we had trained our own brilliance, with which we've created such intrepid robotic explorers, on constructing our human-crafted environment so it worked in harmony with nature, rather than on working so hard to defy natural constraints?

The landscape architect beaver is joined by a host of ingenious animal builders. Take birds; they build a wide array of astonishingly crafted nests. One breed of hummingbird crafts a small structure that looks to all the world like a knob of a tree branch, constructed from bits of bark, shreds of leaves, and spider silk, tied to the branch with more silk, and then covers it with lichen as camouflage to protect its tiny hummingbird chicks. The sociable weaverbird gets its name from the elaborate apartment complexes it builds in tree branches of the Kalahari Desert in Africa. Made of straw, grass, and twigs, these massive constructions, often encompassing the entire tree canopy, house up to two hundred breeding pairs. They look rather like haystacks, but the units are nice and soft inside, lined with cotton and fur. Naturalist Bernd Heinrich writes that they may actually have inspired early humans to begin building domiciles.

Underwater builders can also be ingenious in their design—the master craftsman of underwater architecture being the tiny coral polyp that builds coral reefs, the largest biological structures on Earth, including the Davidson Seamount. And the secret to that little polyp's construction inspired the grandest achievement yet of biomimicry: a recipe for a replacement for concrete. Made from cement combined with sand and rock, concrete is the world's most commonly used building material, and construction's worst greenhouse gas offender. The manufacture of cement alone accounts for an estimated 8 percent of total annual greenhouse gas belching. In fact, it's said that if the cement industry were a country, it would be the third largest national emitter, after China and the U.S. Why? No mere matter of crushed stone slush, cement is made from complex mixes of limestone, shells, chalk, shale, clay, slate, blast-furnace slag, silica sand, and iron ore, which, when heated to 2,700 degrees Fahrenheit, combine to form a rock-hard mass that is then finely ground into cement powder. Copious greenhouse gases are emitted in the process. The coral polyp, by contrast, manages to create concrete-hard calcium carbonate with only the itty-bitty bit of energy that keeps its body humming.

Fortunately, the brilliance of the polyp's reef building snared the attention of scientist and serial entrepreneur Brent Constantz, whose research focuses on biomineralization—the process by which organisms create minerals, like the calcium carbonate that coral polyps manufacture. Constantz founded Calera to commercialize a process of making cement based on his study of corals. He created a means of mimicking the way coral polyps create their exoskeletons, which constitute the concrete-like structures of reefs. The polyp uses carbon dioxide in the ocean; Constantz worked out a way to take carbon dioxide from factory smokestacks and inject it into seawater to create his new type of cement, called Fortera. It's manufacture both sequesters carbon and uses much less carbon-generating heat than traditional cement. That alone was a potentially transformative achievement, but Constantz wasn't stopping there.

Using similar technology, he moved on to found Blue Planet, which produces carbon-sequestering synthetic versions of limestone and sand—two of the biggest components of concrete—also with carbon siphoned from factories. The company's "carbon-negative concrete" has been used in construction at the San Francisco airport, and in 2019 Blue Planet went into partnership with

Kamine Development Corporation, a family-run firm devoted to finding scalable solutions for green construction. Their plan is similar to the Pratt Industries strategy of siting plants near large local supplies, in this case near manufacturing centers producing lots of carbon emissions—in proximity to urban centers, where the most building will be happening. In short, it's a great example of locally concentrated circular production. Constantz says the partnership "marks the point of lift-off for our technology," and, with five thousand sites as first-wave targets around the world for plants, he hopes wide-scale adoption will lead to significant reduction of carbon emissions globally.

Constantz is only one innovator solving the concrete problem. Researchers at Arizona State University have learned how to make a related cement replacement from abalone shells. At the University of Colorado Boulder, another team has created a concrete "that is alive and can even reproduce." It's made by microbes through a photosynthesis process, which also pulls carbon out of the air and turns the cement green, making it look, as researcher Wil Srubar says, "like a Frankenstein material."

Recycling of concrete is also flourishing. Until now, cement couldn't be reused, because it couldn't be extracted from spent concrete. But in 2020, two Dutch companies, New Horizon Urban Mining and Rutte Groep, announced that they'd created a machine that can beat concrete up in such a way that the cement comes free, hence their moniker for their recycled cement, Freement. Used concrete is replacing virgin concrete in new concrete, repurposed for landscaping mulch, and my favorite use, providing the starting structure for new coral reefs.

Growing and restoring reefs is a vital mission, because due to the double whammy of pollution and ocean-water warming, reef scientists estimate that half of the world's reefs have been killed off since 1980. As reefs support about 25 percent of all marine animals, and over 500 million humans also depend on them—not only for the bounty of seafood they nurture, but for prevention of coastal erosion and protection from deadly storm surges—we must do everything we can to protect and cultivate them.

Nature as Inspiration for Structure

When it comes to bringing circular economy principles to the built environment, biomimicry has been particularly influential behind many of the most celebrated and innovative building designs. The iron-and-wood framing of the glass-enclosed Crystal Palace, constructed in London's Hyde Park in 1851 at the direction of Queen Victoria's beloved Prince Albert, was designed by Joseph Paxton, the head gardener of the Duke of Devonshire, based on the webbed support structure of the giant leaves of a breed of lily he cultivated. The interior of Antoni Gaudí's famed Sagrada Família cathedral in Barcelona was meant to invoke looking up from a forest floor, described as "a stone forest of palm trees."

Drawing on nature's architectural wizardry, the ancient Roman aqueducts carried fresh mountain spring water from distances as far as fifty miles to the empire's capital, with an extensive network threaded throughout conquered territories constituting an artificial river system. The Egyptian pyramids are such marvels of gravity defiance that there is still no definitive explanation of how they could possibly have been built to last for so many centuries. Many of the greatest and most lasting structures in history were built with nature's materials: stone, clay, straw, and wood.

Buildings That Scrape the Sky

On Sunday, October 8, 1871, a spark lit the Great Chicago Fire, which raged through 2,100 acres of the city over two days, leaving a hundred thousand people homeless. On October 12, the *Chicago Tribune* ran a story titled "Rebuild the City," which intoned, "All is not lost. Though four hundred million dollars' worth of property has been destroyed, Chicago still exits. She was not a mere collection of stone, and bricks, and lumber. These were but the evidence of the power which produced these things." When Chicagoans rebuilt their beloved city, they embarked on a "conquest over the hostile forces of nature," as one of the city's leading building developers, Henry Ericsson, proclaimed. The result was the advent of the skyscraper, which my friend architect Paul Macht noted to me was the beginning of our current era of horrible energy-consuming, and therefore carbon-emission-producing, building. He pointed out that a confluence of technologies that emerged toward the end of the nineteenth century made constructing buildings that scraped the sky possible.

The electricity-driven elevator, invented by German Werner von Siemens in 1880, allowed for much speedier ascent than the earlier hydraulic models. Electric lighting, made commercially viable starting in 1878 by rivals Thomas Edison and George Westinghouse, and Alexander Graham Bell's telephone, the first model produced in 1877, made conducting work on ever-higher floors practical. Willis Carrier's invention of air-conditioning in 1902 made working in the buildings physically tolerable. But first, the manufacture of superstrong steel beams enabled building higher metal frames for buildings. The former height constraints of enormously heavy brick and concrete walls, topping out at ten stories, were lifted. As Neal Bascomb describes it in his chronicle of the race to the sky, *Higher*, walls became "simply curtains," draped, as it were, over steel. Then those curtains became windows as skyscrapers were wrapped in glass, another brazen act of defiance of nature's wisdom. Paul underscored the inanity; those glass exteriors emit copious heat on cold days and absorb it on hot days, exactly the reverse of what a building should do.

Skyscrapers had plenty of early critics who bemoaned their excesses. While historian Donald L. Miller quotes developer Henry Ericsson saying, "No creation proved more challenging to the human spirit or gave men such a feeling of power and sense of achievement as the spread of two dozen and more skyscrapers around the Loop," not all Chicagoans were so proud. Miller also quotes noted Chicago-based novelist Henry Fuller, who wrote, "In the repelling region of 'skyscrapers,' the abuse of private initiative, the peculiar evil of place and time, has reached its most monumental development."

The chief complaints about building higher in Chicago concerned not only the blocking of sky views and sunlight from sidewalks, but that the buildings were burning so much coal that the city's air was becoming heavy with smoke. Miller observes that skyscrapers had transformed buildings into fossil-fuel-hogging machines. The machine buildings were, nonetheless, highly profitable. In Miller's words, "The skyscraper was as much a machine for turning land into money." Developers drove buildings higher not only to claim bragging rights but to make good on pricey speculation on real estate rights. Every additional floor brought in more income. Regarding New York's epic mimicry of Chicago's skyscraper craze, hometown boy novelist Henry James skewered Manhattan's behemoths as "giants of the mere market" and "mercenary monsters."

The result of the relentless intensification of urban density that ensued is that, as the Ellen MacArthur Foundation's report "Circularity in the Built Environment" highlights, cities now account for 70 percent of global greenhouse gas emissions, consuming 75 percent of Earth's production of natural resources. And of course, we're not nearly done building yet. The report estimates that 70 percent of the buildings India will boast by 2030 are yet to be built, while in China, by 2025, 23 megacities will harbor 5 million residents or more, with 221 other sites bearing populations over 1 million. Brent Constantz cautioned in 2019 that China "has poured more concrete in the past three years than the U.S. had in 100 years." So rapid is the pace of construction that one estimate calculated that between 2015 and 2050, the world will erect a total of 2 trillion square feet of buildings, which equates to building an entire New York City every thirty-five days.

All of which is why a great boom under way in breakthrough circular innovations for construction is so important, facilitating a new race—this time to harmonize with nature. The boom has been long in coming, steadfastly championed by a visionary cadre of green building advocates since the post–World War II population explosion. When I asked Paul Macht to guide me through the important developments, he pointed out the irony that this new wave of innovation is, in many ways, based on a return to the past.

Working with Nature's Ways

Paul and his son, Kyle, who is an architect at Paul's firm, are both specialists in passive solar construction of homes. It has vast potential to reduce emissions, but unfortunately, it's come and gone in waves in the U.S.—starting in the 1930s, seeing a resurgence in the '70s with the oil crisis, but not gaining the traction it deserves. Paul points out that the technology goes back to the ancients. Socrates, in fact, is said to have described the basics twenty-five hundred years ago, as recorded in Xenophon's *Memorabilia*:

> *Now in houses with a south aspect, the sun's rays penetrate into the porticos in winter, but in the summer, the path of the sun is right over our heads and above the roof, so that there is shade. If then this is the best arrangement, we should build the south side loftier to get the winter sun and the north side lower to keep out the winter winds.*

This fundamental wisdom was also well understood, writes architect Dennis Holloway in a wonderful article about the history of the technology, by ancient Native American builders, who followed the principles in constructing such ancient sites as Mesa Verde in Colorado, Grand Gulch in Utah, and Chaco Canyon in Arizona. Yet, despite reducing energy use by as much as 75 percent, he observes, most houses built in the U.S., "even in the Sun Belt states, don't make any economic use of the sun's energy."

Holloway points out that one reason is that interest in passive solar was beaten back during the Reagan presidency. "There's a stubbornly persistent myth," he writes, "a holdover from the news media coverage of some of the early passive houses during the first Reagan Administration, that overheating in summer is common.... This is a propagandistic deception by Big Oil." Today, there can be no doubt that the technology is superb, with much learned about siting of buildings, optimal materials for absorbing heat during the day and releasing it at night, and best placement of ventilation to lower, or eliminate, the need for heating and air conditioning. The Intergovernmental Panel on Climate Change (IPCC) identifies reducing use of refrigerants as the number one action for addressing climate change. It's becoming all the more urgent as more of the developing world adopts air conditioners, and as temperatures rise.

Another return to the past is a resurgence in the use of natural materials, to displace man-made carbon offenders. One is good, old-fashioned wood. Building with wood is making a big comeback all around the world. Reclaiming wood from buildings being destructed is one best practice. Using wood from trees on a property is another great option. Paul and Kyle used only wood from cherry trees on the property of the Pocono house they built. Using wood for siding is also coming back, due to a new treatment called thermal modification, heating the wood to 400 degrees Fahrenheit or more. That makes it water repellent as well as more resistant to incursion by insects, like termites, as well as to decay. As Paul describes the process, "The food source is cooked out of the wood."

Building with more wood is part of what Kyle and Paul call "the next big thing" in green building—cutting down on embodied carbon. As opposed to the greenhouse gas emissions due to the ongoing operations of a building, which is "operational carbon," embodied carbon refers to all the emissions involved in the making of the building, from the energy used to make steel beams and concrete, to the transportation of materials and the building process itself. It can account for 20 to 50 percent of a building's carbon footprint, and Paul highlights that "cutting embodied carbon emissions of buildings being constructed is a one-time huge hit that's right now."

Wood is a carbon storer, for as long as it's not burned or otherwise degraded, so using it as the primary structural material can make buildings carbon sinks. A study conducted at the Yale School of the Environment, by a team of researchers led by Galina Churkina and including ecosystem services pioneer Thomas Graedel, argued that constructing mid-rise buildings in cities with mass timber "has the potential to create a vast 'bank vault' that can store within these buildings 10 to 68 million tons of carbon annually that might otherwise be released to the atmosphere."

Wood has historically had a natural limit of only about six stories, but no more—the new wave is much taller. I said good, old-fashioned wood is making a comeback, and that's true, but as far as tall buildings go, it's newfangled wood that's making them possible. A number of forms of reengineered wood have been created that are much stronger, referred to collectively as "mass timber," with the most commonly used type being cross-laminated. Lumber boards are glued to one another in layers, in a crosswise pattern, with the grain going against the grain of each succeeding layer, to make slabs, rather like concrete slabs, that are a foot thick.

That exciting potential is why a steady stream of architects, developers, and city planners have been making their way to the little rural Norwegian town of Brumunddal, population 8,700. An 18-story, 280-foot (85-meter) wooden apartment complex opened there in 2019 that's been certified as the world's tallest timber building. All of its supporting columns and beams, its facing, and even the elevator shafts, are made from laminated wood, grown only miles away in a sustainably managed forest. Even better, the building developers are planting two trees for every one felled for the construction. In another intriguing feat of biomimicry, the wood for this building was configured in a fire-resistant pattern, based on a study of why large logs thrown on a campfire need lots of kindling in order to burn.

Numerous plans have been drafted for even taller wooden sky castles, such as the seventy-story timber W350 Tower to be built by Sumitomo Forestry in Tokyo. But the biggest push now, Paul says, is for five- to seven-story cross-laminated buildings. Should the practice truly take off, it would not only be a marvelous means of sequestering many tons of carbon, but could be another major boon to reforestation—*if* strict sustainable forestry standards are met.

Many other natural materials that were long-ago mainstays of construction are being brought back. Paul and Kyle are fans of waste straw from farming operations, which is being used for insulation, a stellar alternative to most current versions, which are primarily made of plastic. "There's enough waste straw in the U.S.," Kyle says, "to insulate all the houses in the country." Their preferred insulation, though, is made of cellulose, from recycled paper. Both are great for another building technique they advocate: superinsulated construction, an outgrowth of passive solar building. It involves much thicker insulation than is typical, and windows are ideally triple pane and double insulated.

Paul and Kyle see the move to natural materials as part of growing awareness of the toxicity of the materials commonly used in building construction. A set of standards called the Red List has been highly influential, offering a long list of harmful chemicals that must be excluded from any materials in order to meet the standards of the Living Building Challenge.

Architect Jason McLennan threw down a new gauntlet to the building industry when he wrote the Living Building Challenge, a set of sixteen extremely demanding requirements for advancing John Lyle's concept of living buildings. To meet the standard, buildings must not only be energy efficient but net-positive clean energy producers, as well as water cleansing, air purifying, habitat providing, and beautiful. McLennan is the director of the International Living Future Institute, which promotes adoption of the standards. In a nod to biomimicry, which has been a keen interest of McLennan's, he begins the Living Building Challenge document this way: "Imagine a building designed and constructed to function as elegantly and efficiently as a flower: a building informed by its bioregion's characteristics, that generates all of its own energy with renewable resources, captures and treats all of its water, and that operates efficiently and for maximum beauty."

The analogy is not merely a metaphor. The standards are divided into six "petal" groups—materials, site, water, energy, health and happiness, and equity and beauty, with achievement of all the standards required for Petal Certification. That stipulates that the building must use only the amount of water that can be harvested on-site, primarily through rainwater capture, and must purify it in a closed-loop recycling system. Buildings must produce 105 percent of the energy they need for operations, and all materials must not violate the Red List. Certification is not awarded until a building has been operating for twelve months and has proven it's made the grade. First issued in 2006 and updated several times since, the challenge has sparked a new race for building bragging rights.

I spoke to McLennan about his goals for the challenge. He's quite philosophical, and a heartening optimist, but he's also tough minded and not at all afraid to call his industry out on its bad behavior. "Nature—not ego and fashion—dictates the parameters of a green building," he writes in his book *Zugunruhe*. The title of the book is the German term for the "migratory restlessness" birds feel in the days before they take wing, as they feel the air cooling. McLennan says he senses a comparable urgency to get moving among his colleagues in the building trade, to move further and faster into the greener world he's working to foster. In this case, of course, due to atmospheric warming.

Birds have made a profound impression on McLennan. He recalls his wonder as a child in learning that tiny hummingbirds flew from his home in Sudbury, Ontario, to winter haunts in Mexico, fifteen hundred miles away. Masters of long-distance aviation as well as crafters of ingenious camouflage! We all must make our own inner migration, McLennan reflects, to a new consciousness, if we're to heal our planet.

His own consciousness was transformed by working with the

renowned architect Robert Berkebile, founding partner of one of the most influential green building design firms, Berkebile Nelson Immenschuh McDowell (BNIM), which opened its doors in 1970. McLennan joined the firm straight out of architecture school. Berkebile was a leading voice in championing the Leadership in Energy and Environmental Design (LEED) building standards released in 1998, which have set the bar for green building worldwide over the past twenty years.

Berkebile and a group of fellow visionaries, including National Resources Defense Council scientist Rob Watson, known as the Founding Father of LEED, were driven to create LEED because they perceived that without rigorous standards for calling a building green, developers had little incentive to invest in incorporating a host of green innovations that had been pioneered in the prior decades. Some truly stunning achievements had been unveiled, to much oohing and aahing but far too little ripple effect.

William McDonough of the Cradle to Cradle Institute practiced principles of what he called "ecologically intelligent design" in his first large commission, the Environmental Defense Fund's new headquarters building in New York City, which opened in 1985. German architect Rolf Disch blazed the way for net-positive energy buildings with his Heliotrope house, completed in 1994, which was the first building in the world to generate more energy than it uses, rotating throughout the day so its solar panels can track the sun. An astonishing breakthrough of building biomimicry was the Eastgate Centre, a 350,000-square-foot combination retail and office building in Harare, Zimbabwe, that opened in 1996. The building was designed by Zimbabwean architect Mick Pearce according to his discoveries about how termites in Africa regulate the temperate inside the enormous mounds—up to 30 feet high—they build. He found that the termites stabilize the heat at about 87 degrees Fahrenheit, perfect for the growth of a fungus inside the mounds that's their main food source. They do so by constantly digging or plugging up vents in the mound's walls in order to create a flow of air that brings cooler air up from the bottom and circulates it. Mimicking this system with fans and vents, the East-gate Centre has no heating or air-conditioning units, running at 10 percent of the energy use of a typical building of its size.

But such advances were much too few and far between. The LEED standards did accelerate adoption of many important means of building greening, such as environmentally sensitive siting of buildings and better energy efficiency. But a flaw is that buildings are awarded points for attaining either Platinum, Gold, or Silver certification in one or a few of those areas while they might fall far short in others. The result, as one architectural historian writes, has been "LEED brain," with developers "driven by earning points rather than truly by sustainability." Robert Berkebile himself points to another flaw—that the standards set the bar too low. "The certification has become: Your building is doing a little less damage to the environment than everyone else's," he says "I think that's a failure."

Demonstrating that conviction, Berkebile helped set the pace of adoption of the Living Building Challenge criteria, designing one of the first two buildings to be christened officially living: the home of the Omega Center for Sustainable Living in Rhinebeck, New York, which opened in 2009. There are now five hundred buildings around the world pursuing LBC certification. That may not sound like many, but recall the impact just a couple dozen early skyscrapers in Chicago had.

The challenge is also catalyzing a green materials revolution. "So many manufacturers want to get their products to meet the Red List standards," Paul says, referring to things like flooring, insulation, and wallboard. "It's exciting that when we talk to firms about their products, which we do all the time, they're getting on-board." Most people have no idea, he cautions, how many toxic chemicals may be built into their homes. After all, "people can't see what's inside their walls." As of now, 90 to 95 percent of single-family homes built in the U.S. are built by large corporate developers, he stresses, and their basic rule is "build as cheap as you can." That involves using various forms of plastic and other synthetics that not only can release toxins but are also carbon intensive. "Our homes," Paul says, "should be as healthy for us as our food should be."

Some pushback with any big leap of standards is inevitable, but Jason McLennan argues there is no good reason for architects and developers not to adopt the challenge goals, not even economics. "Every time industry screams that higher standards will bankrupt them—the opposite has proved true." Indeed, the buildings thus far constructed to the Living Building standards are enjoying great success. The solutions they've devised are wonderfully diverse and deeply circular.

The Bullitt Center building in downtown Seattle was commissioned by Dennis Hayes, Earth Day founder and now director of the Bullitt Foundation, which promotes sustainable urban development in the Pacific Northwest. It's proven, with its solar-paneled roof, that even in a far northern, often rainy city, solar power can reliably produce net-positive energy for a large commercial building. But what I'm most intrigued by in the building is its water system. On a green roof built into the third floor, a wetland has been cultivated that filters water harvested from rain in a repeating loop, assuring the plants in the wetland receive a good dose of natural nutrients. The Kendeda Building for Innovative Sustainable Design, on the Georgia Tech campus, opened in December 2019, and before the COVID-19 pandemic hit, it was exceeding predicted energy production for three months in a row. I was impressed to see that it was built by one of the world's largest and most influential construction firms, Skanska, which is a good indication that the truly big boys are getting in on the action.

McLennan sums up the progress made so far this way: "The models have to be pragmatic and beautiful. They have to make sense. This is the phase we're in now. We're all in a bit of a hurry to show the world that this different way of building is better, to build the deep green examples.

10

Scaling Circularity Up

F IVE YEARS FROM NOW, at 7:00 a.m. on Monday, another day
of circular living begins . . .

You're gently woken up by your five-year-old modular
smartphone that works like new because it can continually be refurbished and upgraded. You feel refreshed waking up on sheets
with the feel of a luxurious fabric—because they were manufactured from nutrient-rich algae. Stepping into the shower, you pump
out body soap from your stylish, 100 percent recycled refillable
shower soap dispenser. You get all your cleaning products as refillables now, paying only once for high-grade packaging when
you first buy the product; any disposable bottles you use are either
manufactured with recycled material and are 100 percent recyclable or are made of plant-based materials that are compostable.
Heading to the kitchen, you fire up your gas range, powered entirely by your anaerobic digester that also heats all your water. You
haven't paid a gas bill since you installed the digester three years

ago. You rarely have to take out the garbage, and disposal costs to your home and community are negligible.

Grabbing the milk out of the fridge for your cereal, you notice that the color indicator on the front of the carton is verging from green to yellow, warning you the milk will be going bad in two days. With so many food companies affixing color-sensor warning patches to their food packaging, you can't remember the last time you discarded food because you were unsure of its freshness. You pour some milk on corn flakes you brought home from the store in your reusable container. The flakes were made from corn grown through regenerative farming. A boom in regenerative farming around the world has already drawn down billions of tons of carbon from the atmosphere, reducing the effects of global warming.

Your clothes are made from circular and nutrient-rich material. Your skin has never felt better. Hopping into your fully charged electric SUV, powered by your rooftop solar panels, which also provide all your household electricity, you head to your office in a Living Building, constructed from a carbon-sequestering cement replacement. You feel that every day, as more and more ingenious circular economy innovations come to market, you're doing your part to combat the climate problem.

THIS SCENARIO OF CIRCULAR LIVING is entirely plausible. As the wealth of innovations introduced in the preceding chapters have shown, we have the knowledge and the technology available to make all these products and services, to dramatically eliminate waste, and to build our homes and offices so that they renew the health of the environment.

The question now is, how can we accelerate progress? How can we scale up proven solutions?

One answer is by using our individual buying power to support circular start-ups and the major consumer goods companies introducing circular products and packaging to the market. The large consumer goods companies are in a particularly strong position to innovate and create markets for recycled, sustainably grown, and biodegradable materials. They have deep expertise in product development to bring to bear, as well as financial and market clout to throw behind transitioning. In contemplating the best means of accelerating the transition to circularity, I spoke with two leading figures in promoting circular consumer goods, Seventh Generation founder Jeffrey Hollender and former CEO of Unilever Paul Polman.

Recall that Hollender is also the head of the American Sustainable Business Council, which he founded. He is nothing if not a fighter; launching a bold new push for corporate reform when he could be resting on his laurels. He scaled Seventh Generation from a badly struggling start-up friends thought he was mad to keep plugging away at into a category-leading brand, which has catalyzed the development of green cleaning products throughout the industry. With its acquisition by Dutch consumer goods giant Unilever in 2016, it became a global force with a powerful presence in forty countries. In short, he knows a thing or two about scaling.

I spoke with him from his home on beautiful Lake Champlain in Burlington, Vermont, home to Seventh Generation's headquarters. Hollender exemplifies the tenacity and depth of commitment that will be required of business leaders to drive transition within their firms. He seems to be a natural iconoclast, having attended the unconventional Hampshire College, which gives no grades, and leaving before graduating to pursue his own uncharted path of self-learning, which included helping others who couldn't afford college to learn. His first business, founded in Toronto when he was twenty-two, was education nonprofit the Skills Exchange, an innovative low-cost provider of short, practical classes taught by professionals in their own offices or homes. He was inspired, in part, by auditing a class by Marshall McLuhan, then teaching at the University of Toronto. McLuhan was working on his book *City as Classroom*, which argued that most learning is done outside of classrooms. Hollender had just shown up at McLuhan's class one day and asked if he could attend.

He showed the same seize-the-day initiative in building Seventh Generation. Growing up in the roiling 1970s, as the environmental movement exposed how toxic so many products and production processes were, he wanted to be part of finding solutions. His participation in Vietnam War protests had taught him, as he said, "If you cared about something, you had to show up to do something." When Hollender perceived an opportunity to create healthier and more planet-friendly cleaning products, Seventh Generation was born. It did not grow up without struggle.

"Except for one year," he told me, "we lost money for thirteen years. I had the support of my immediate family, but I had friends who thought I was crazy. They'd say, 'Why don't you pronounce this dead and move on?' But I am not someone who gives up." Jefferey had the absolute conviction of an entrepreneur on a mission to change the world. His patience and persistence paid off in both impact and returns to his original investors after Unilever acquired Seventh Generation for over $600 million.

The turning point in the company's growth came when Whole Foods, which was itself growing dramatically, began selling their products and devoted a whole aisle to Seventh Generation products in some stores. That speaks powerfully to another force for scaling: the synergies between the benefits of circularity for brands and retailers. Whole Foods was creating a green-products ecosystem, which, as Hollender put it, created "trust in Seventh Generation because of the aura Whole Foods had." Amazon, as the world's largest retailer and having bought Whole Foods as a step in that direction, can and should use its enormous market power to greatly grow that ecosystem. Walmart, Costco, Target, all the world's leading retailers have the potential, and the incentive, to considerably expand it as well. Sales of consumer goods marketed as environmentally friendly have increased at a much faster rate in recent years than those that aren't, with a New York University study showing that between 2015 and 2019 they accounted for more than half of the total growth in sales.

When it comes to levers to prod business leaders to more vigorously respond to the demand, one approach Hollender is a big fan of is true cost accounting. Only a tiny sliver of companies has performed calculations to estimate their carbon footprint, let alone the harm they're doing to the planet in myriad other ways. Even fewer have made those figures public. The good news is that true cost accounting has come of age, thanks largely to the pioneering commitments of the likes of Puma's Jochen Zeitz. But one impediment to large-scale adoption is that multiple methods have emerged.

In addition to the approach of consulting firm Trucost, who worked with Zeitz, Stanford professor Gretchen Daily has championed the Natural Capital Project, in partnership with the World Wildlife Fund, which provides open-source software for true costing; in Europe, the International Organization for Standardization has issued analytic tools for companies. A push should be made to merge methods into one internationally agreed standard and to provide low- to no-cost access to tools to do the number crunching. But that's just the beginning. As Hollender and the Sustainable Business Council argue, companies should be required to publicly report on their environmental impact and consumption of natural capital throughout their supply chains, and to compensate the public for the taxpayer money required to address the harm they've done. They've been free riders for far too long. Meanwhile, the longer they continue polluting, pumping their products with toxins, and paying punishing wages, the more vulnerable they are to public exposure and censure.

Hollender told me he learned the lesson of transparency the hard way. His team had worked intensively to make their laundry detergent free of toxins, but they couldn't figure out how to rid the liquid of a chemical by-product called a surfactant, which has adverse health effects. Although their detergent had much less than most, they hadn't disclosed it as an ingredient, and the brand took a hit when the Organic Consumers Association tested a hundred products and found this contaminant in all of them, including Seventh Generation's. "It's one thing to make a compromise and tell people about it," he told me, "but what really violated our values was making a compromise and not being transparent about it." Seventh Generation went on to find an alternative surfactant and to join the EPA's Safer Detergents Stewardship Initiative to help solve the problem.

As Hollender found, as had Jochen Zeitz at Puma, being open about the environmental damage a company is inflicting, as part of a push to constantly do better, builds bonds of trust with customers as well as investors. It also announces to innovators, "We are interested in your innovations. Come to us first." Neither consumers, nor market analysts, are naive anymore about what Zeitz called "bijou" green offerings that are truly just greenwashing.

A bold move by Unilever in 2020 shows signs of building market pressure for transparency. The company announced that it will gather and publish carbon-footprint data from all its suppliers, after the company's transformative former CEO Paul Polman, who stepped down in 2019, built a lasting culture of belief in his pioneering environmental and economic justice initiatives, which he dubbed the Sustainable Living Plan.

During his ten years at the helm, Polman made a series of acquisitions of brands making serious commitments to improving environmental and social conditions, including, in addition to Seventh Generation, Ben & Jerry's, Dove, Lipton, Hellmann's, and Knorr. In announcing the plan, Polman argued that "sustainability isn't just the right thing to do, it is essential to drive business growth," and while he faced considerable pushback from board members and powerful shareholders, he handily made the case. In 2019, Unilever announced that the twenty-eight total Sustainable Living brands grew 69 percent faster than its other brands and accounted for 75 percent of the company's growth. The market has most definitely taken note—Unilever delivered a whopping 290 percent return on shareholder investment over the course of Polman's tenure. The company also reports that it avoided $1 billion in costs. Polman stresses that becoming a good environmental steward is vital for attracting the best workers, as so many millennials consider a company's commitment to environmentalism as a major factor in whether or not to accept a job offer.

While Hollender steadfastly built sustainability into his business from the bottom up, Polman relentlessly imposed it from the top down. His initiatives went far beyond buying up businesses that had led the way. He drilled deep down into the issues with suppliers, promoting sustainable farming, fighting against deforestation, setting a goal of using 100 percent renewable energy for its production, and providing protections and economic opportunities for women. At each step, he dealt with skeptics. "When I started saying seven or eight years ago," he told me, "that I wanted all of our factories to be zero waste to landfill, people thought I was loony. They thought that was going to cost lots of money, because there's this notion that if you do the right thing, it must cost a lot of money. The opposite is true. Green energy is now cheaper than fossil. Much of the world is now on carbon pricing or cap and trade." Unilever didn't have problems with those regulations because it had already gotten ahead of the game. He said he never saw a payout of longer than three years for green investments.

"When I invested twenty to thirty million dollars in a tea plantation," he told me, "to give the employees decent housing and give them proper sanitation, and planted drought-resistant tea bushes, people probably thought I was nuts, but now it's the highest yielding tea plantation in Kenya." He set a goal of working with five hundred smallholder farms for a range of produce, with the requirement that half of them be owned by women. He investigated situations on the ground to get the full picture. "You have to go deep," he said. "You have to be there to see it."

When I asked him where his drive to constantly push the envelope of sustainability came from, he revealed that his first job, at Procter & Gamble in 1979, was taking waste out of the system. "So much of the waste we're creating," he said, "we're creating just because we are not thinking properly." His upbringing also pointed him on the way. He grew up in Holland, born just eleven years after World War II had devastated the country, and his parents put a premium on helping other people. He initially wanted to become a priest but wasn't admitted to seminary; then he aimed at becoming a doctor, but also didn't get into a program. Then, he says, "I discovered in business, you can have more impact."

You can, if you will. "So many CEOs prefer to be on the golf course," Polman said, and "so many companies have negative externalities that they expect the government to cover for them. I just don't want to be a part of that. I don't want to step on others' toes, or cut off other people's opportunities, or make others' lives miserable." And all along, he showed that there was absolutely no business need to do so. Unilever even paid 24 to 25 percent in taxes year to year, he reported.

For years, incredulous business leaders would ask him, "Why do you have so much consumer trust? You must have a good PR department." The answer was that Unilever was listening to their customers and offering products that met their evolving needs, as well as increasing focus on transparency as it relates to the social, environmental, and governance characteristics of a brand. Polman added that he sees a change in recent years, which he attributes to the financial market taking these commitments seriously, and says a "bifurcation is happening in the market between industries of the past and industries of the future." Hollender, Polman, Jochen Zeitz, Yvon Chouinard, and the green initiatives of Walmart, IKEA, Dell, and Hewlett-Packard have solidified the business case, and the growing market preference for authentically responsible stewards of environmental and human well-being is perhaps the greatest force for scaling up.

Persuasive as the arguments based on environmental degradation should be, it's emphasizing the economic opportunity that's more likely to advance the cause among business leaders most vigorously. The World Economic Forum estimates that developing circularity in just the consumer goods sectors could yield annual savings of $700 billion. A McKinsey analysis concludes that in the apparel industry, $500 billion in losses could be recouped annually—and that's not accounting for the environmental benefits. Recall that the overall economic payoff of implementing entirely reasonable circular processes by 2030, as calculated by Accenture, could be as much as $4.5 trillion. The favor of major investors is another opportunity business leaders will find increasingly tantalizing.

Evidence has been mounting fast and furiously that the most sustainable companies are also yielding the highest investment returns. A Harvard study that analyzed the results of 180 U.S. companies determined that those they deemed "high sustainability" firms "significantly outperform their counterparts over the long-term, both in terms of stock market as well as accounting performance." Research by financial assets management firm Arabesque revealed that companies ranked in the top fourth of the S&P 500 for responsible environmental, social, and governance practices (ESG) outperformed those in the lowest fourth by 25 percent on earnings, with less volatile stock prices. Leading investment managers are taking note. Microsoft recently announced it was creating a $1 billion Climate Innovation Fund, to accelerate the development of carbon emissions reduction and drawdown technologies, and Amazon doubled that commitment, forming a $2 billion venture capital firm to invest in carbon reduction and other climate remediation technologies.

With a patient approach, the remarkable economic energy of the private sector can be the most powerful driver of change. While the Industrial Revolution generated enormous wealth as well as important innovations, its systems are often pollutive to the point of killing the consumer and land that it relies on for revenue and material. In the Circular Economy revolution, the returns to investors will be much greater because its practices are sustainable, aligned with the ecosystem of resources and customers on which companies and economies ultimately rely on for profits and growth.

Ever More Circular Cities

Growing up as a track star in Phoenix, Ginger Spencer was so determined to constantly improve her performance that she practiced with her high school men's team. The first member of her family to attend college, she got a master's degree in public policy from Carnegie Mellon. She took a fascinating class called Management Science, which was so notoriously tough that the professor told the students half of them wouldn't even finish the final exam, let alone pass it. He also told them, she recalls, "You are going to be presented with challenges that seem impossible to solve, where there is no answer, but if you dig deep enough, you will find a solution." She's brought that same drive for excellence and dig-deep problem-solving ethic to her work as the director of public works in Phoenix, with remarkable results.

Phoenix was dubbed the world's least sustainable city in 2011 by New York University professor Andrew Ross in his book *Bird on Fire*, a scathing exposé of the environmental degradation caused by building a sprawling, thousand-square-mile city with a fast-growing population of 1,700,000 in a scorching desert. Ross said Phoenix was in "the bull's-eye of global warming," and argued that attempts to make the city more sustainable would be an acid test for how effectively the world will be able to tackle climate change. Ginger Spencer and her team have made great progress, a testament that with the right political leadership, cities can quickly implement major circular initiatives.

That political leadership came from Greg Stanton, elected mayor in 2012, the year after Ross's book was published—and it became a rallying cry for the sustainability platform he ran on. The public heard the call. By 2017, Phoenix was named one of the ten best cities around the world for taking climate action by the Bloomberg Philanthropies C40 Cities group, a network of forty of the world's largest cities that have committed to climate change remediation. The approach the city has taken should serve as a model around the globe, as Ginger's team has collaborated with professors at Arizona State University to conduct research on best prospects to implement and to assess the success of programs. For one of those, a new composting venture, they verified that the compost created by the city and used to organically green the city's parks, is a better fertilizer than chemical options. The emphasis is on constant innovation, and the city put out a call to anyone around the world to suggest solutions.

One idea submitted was to turn the city's masses of palm fronds, from the palm trees that grow in virtually every yard, into animal feed for the many small ranches that still ring the city. Ginger and her team snapped into action, making a deal with a California-based company, Palm Silage, to create a facility in Phoenix. Unfortunately, after three years, the program wasn't generating adequate revenue and had to be discontinued. But Ginger is undaunted. "We're learning lessons as we go," she told me, and she's pressing ahead on other fronts, like her arrangement with Renewlogy, an innovative early-stage company that is converting hard-to-recycle plastics back to basic molecules to be rebuilt into new plastics. When National Sword struck, and the city was faced with the need to increase the rate it charges for recycling pickup, she led a full-court press public campaign to build support, sending members of her team door to door throughout the city, holding two to three meetings a day with community groups and elected officials, launching a website, and conducting a citywide survey. The rate increase was passed with 80 percent public support. With such determined and creative leadership, cities can be crucial hubs for accelerating circular innovation. And in addition to the C40, which includes most of the largest cities in the world, some smaller cities are boldly pioneering.

In England, the political and business leaders of the city of Peterborough, with a population just over 200,000, came together to declare in 2016 that the city would become fully circular by 2050. They've since developed a detailed road map for transformation, which includes an industrial ecology venture, already up and running, that collects surplus bread from all around the city and turns it into beer at the Baker's Dozen Brewery. Other plans include the installation of sensors throughout the city to monitor air quality and carbon emissions, goals to achieve zero carbon energy use, and regulations for building with sustainable materials and for eliminating plastic use. In the U.S., leaders in Charlotte, North Carolina, issued the Circulate Charlotte economic development plan that aims to make the city zero waste. It includes developing the local agriculture industry and farm-to-table food-supply chain; creating a smart, solar-powered energy grid; and even creating aquaponic fish farms in the public schools to provide fresh fish for schoolchildren, along with a circular economy curriculum.

Circularity Going National and International

Political leaders at the national level are instituting initiatives and passing legislation that will accelerate the transition to a Circular Economy. The European Union has been a bold leader, and in 2018 the EU and China signed a Memorandum of Understanding on Circular Economy Cooperation, which pledges each to engage in ongoing dialogue about how best to advance circular initiatives. Europe is well out ahead, with the European Union having agreed in 2015 on its Action Plan for the Circular Economy, updated in 2020, and proposing legislation that would set strict targets for manufacturers' use of recycled materials and impose tough restrictions on single-use plastic. France has already outright banned plastic plates and cups. In England, the national government announced a strict set of waste regulations that will require producers and retailers of plastics and food waste to pay the entire cost of collection and recycling, and include fines for producing particularly problematic materials, like black plastic trays.

In response to the COVID-19 pandemic, the EU nations agreed on the largest economic package directed at climate change to that date, pledging over 500 billion euros to support the adoption of electric cars, development of renewable energy and carbon capture technologies, expansion of public transportation, and installation of green technology by homeowners, among other solutions.

China has also announced ambitious new targets for lowering emissions, asserting it is seeking carbon neutrality by 2060, and aiming for 84 percent of its energy consumption by that date to be of renewables versus only an estimated 20 percent in 2025.

We're making significant progress in North America too, with Canada leading the way. The province of Ontario passed the Resource Recovery and Circular Economy Act in 2016, which instituted producer responsibility for the collection and recycling of waste. It was updated in 2020 to include new standards for making plastic packaging recyclable and for developing the infrastructure for turning waste into energy. In the U.S., in addition to right-to-repair laws introduced in twenty states, Senator Tom Udall and Congressman Alan Lowenthal have proposed the Break Free from Plastic Pollution Act. Jeffrey Hollender's American Sustainable Business Council is putting forth a detailed set of policy proposals it will lobby for, which include establishing full-cost accounting requirements, transparent reporting by companies of their materials sourcing, and the shifting of subsidies from fossil fuel companies to renewables innovators.

CONCLUSION

The Future Is Circular

A S WE MOVE INTO the next evolution of product design and manufacturing, the circular economy will provide us the opportunity to benefit from the innovations of past revolutions in manufacturing while also maintaining our personal, family, and societal health. Transparency will be at the core of production, which will lead to the equitable practice of ensuring that manufacturers and brands will profit or lose based on their ability to generate holistic value—that includes their shareholders, customers, and the communities in which they manufacture and sell. Those that profited in the previous era will be consigned to their rightful place: the wastebin of history. Those that build value from circular business practices, aligned with their shareholders, customers, and their communities, will build lasting value in a world in which building a profitable business and preserving our health and environment are synonymous. In doing so, we can repudiate the gospel of waste and instead champion the ethic of circular consumption.

If companies want to participate in enjoying the benefits of a capitalist system, they can no longer expect the public to manage the health effects of their pollution or subsidize the disposal of their products with tax dollars. Multiple companies have already demonstrated the economic value created by aligning the interests of shareholders, employees, customers, and our environment: Patagonia, in the fashion industry; Seventh Generation in the cleaning products industry; Ben & Jerry's in the food industry; SodaStream in the beverage industry; Cascade Engineering in the manufacturing industry; and Phillips in the equipment industry are just a few examples.

Perhaps the best example of a leader who has propagated such economic value is Paul Polman at Unilever. During his ten-year tenure as CEO, Polman was a relentless champion of sustainable business practices, leading the company to acquire some of the most sustainable consumer product companies, all while challenging traditional brands already in the Unilever portfolio to become more sustainable. In business, results matter. During Polman's tenure, Unilever delivered a total shareholder return of 290 percent, far surpassing the returns generated by its competitors.

Most emblematic of the opposing view on how a business should conduct itself is Kraft Heinz. As Polman recalls, when they attempted a hostile takeover of Unilever, "It was a purely financial transaction that was attractive on paper, but was really two conflicting economic systems. Unilever is a company that works for the long term and focuses on the billions of people that we serve. Kraft Heinz is clearly focused on a few billionaires who do extremely well, but the company is on the bottom of the human rights indexes or on the efforts to get out of deforestation. Kraft Heinz is built on the concept of cutting cost. Since then, their share price is down 70 percent, and they now face legal issues around reporting. Our share price is up about fifty percent. Some people think greed is good. But over and over it's proven that ultimately generosity is better."

With the data clearly showing that businesses that align the interests of shareholders, employees, customers, and our environment create a winning strategy for longevity and value creation. But it is shocking how many CEOs remain focused simply on their quarterly earnings, and how many investment banks still invest only a small portion of their funds in a way that prizes strong social, environmental, and governance practices.

One of the worst corporate citizens of the past one hundred years, for example, is ExxonMobil. They hid information from the public about the harmful effects of their supply chain and product while donating huge sums of money to unscrupulous politicians in order to gain financial subsidies extracted from the same public that they were harming. When alternatives such as solar and wind came to the market, they didn't behave like the capitalists they purported to be and invest in the alternatives, or decide to fairly compete on consumer preference and price. Instead, they used duplicitous marketing tactics to try to discredit those alternatives. The result? ExxonMobil's stock declined by over 50 percent in five years. It now faces the dire scenario in which much of its assets—oil fields and pipelines—are being considered "stranded assets" as the cost of alternative energy continues to decrease and demand increases. Unfortunately, the lobbying of underhanded politicians, the living off of public subsidies, and the deceptive marketing tactics have allowed executives at ExxonMobil to retire with millions of dollars, leaving a lasting legacy of pollution that all of us will have to share in the cost of cleaning up for decades.

ExxonMobil is not alone. The list of companies that have caused comparable damage by employing such shady methods to avoid competition and transparency is long. But the good news is that there is hope, there is inspiration, there is leadership, and there is a movement that includes multiple stakeholders. Brilliant innovations in materials science, product design, supply-chain transparency, advanced manufacturing technologies, and sustainable, high-quality consumer products and services are creating a circular economy that will create equitable and lasting value. CEOs like Paul Polman and Yvon Chouinard, investors like George Soros and Jeremy Grantham, political leaders like Angela Merkel and Jacinda Ardern, and, most important, every one of us, are leaders in this transition to a circular economy.

Acknowledgments

I AM FORTUNATE TO have friends, teachers, coaches, mentors who have each selflessly provided soil to the hill I stand on to look beyond the horizon and see the destination that has become this book. I am grateful to the people who have joined me at my various ventures through the years to help make my ideas a reality, with a special thank you to Bob, Bridget, and Alicia.

Notes

Introduction

ix **Once a vibrant wetland:** Ted Steinberg, *Gotham Unbound: The Ecological History of Greater New York* (New York: Simon & Schuster, 2014), 248.

ix **When the area was selected:** Steinberg, *Gotham Unbound*, 247.

ix **As feared, the 2,200 acres:** Nancy Reckler, "New Yorkers Near World's Largest Landfill Say City Dumps on Them," *Washington Post*, August 7, 1996.

x **Rounding a bend:** William Bryant Logan, "Lessons of a Hideous Forest," *New York Times*, July 21, 2019.

xii **Americans currently recycle:** ISRI 2019 Annual Report, Institute of Scrap Recycling Industries, Inc., Washington, D.C.

xii **What's more, they would forgo:** ISRI 2019 Annual Report.

xiii **But an estimated two thirds:** Circle Economy, The Circularity Gap Report 2019, Platform for Accelerating the Circular Economy (PACE), 24. https://circulareconomy.europa.eu/platform/sites/default/files/circularity_gap_report_2019.pdf.

xiv **More frequent and severe:** McGrath, "Climate Change."

xiv **Rain forests, which are the most powerful:** Rhett A. Butler, "10 Rainforest Facts for 2020," Mongabay.com, July 12, 2020, https://rainforests.mongabay.com/facts/rainforest-facts.html.

xiv **Research shows that:** Henry Fountain, "Climate Change Is Accelerating, Bringing World 'Dangerously Close' to Irreversible Change," *New*

219

York Times, December 4, 2019, https://www.nytimes.com/2019/12/04
/climate/climate-change-acceleration.html.

xiv **The United Nations estimates that:** United Nations, UN Water, "Water Scarcity," accessed October 1, 2020, https://www.unwater.org/water
-facts/scarcity.

xiv **Fourth National Climate Assessment:** Carmin Chappell, "Climate Change in the US Will Hurt Poor People the Most, According to a Bombshell Federal Report," CNBC, November 26, 2018, https://www
.cnbc.com/2018/11/26/climate-change-will-hurt-poor-people-the
-most-federal-report.html.

xv **For example, *Fortune* reported:** Adrien Salazar and Lennox Yearwood Jr., "The Next President Must Tackle the Intertwined Crises of Racial Injustice and Climate Change," *Fortune*, October 10, 2020. https://fortune.com/2020/10/10/trump-biden-climate-change-racism
-hurricane-delta.

xv **As for indigenous peoples:** United Nations Permanent Forum on Indigenous Issues, "Climate Change and Indigenous Peoples," accessed January 1, 2021, https://www.un.org/en/events/indigenousday
/pdf/Backgrounder_ClimateChange_FINAL.pdf.

xv **A third of Earth's soil:** Chris Arsenault, "Only 60 Years of Farming Left If Soil Degradation Continues," *Scientific American*, December 5, 2014, https://www.scientificamerican.com/article/only-60-years-of
-farming-left-if-soil-degradation-continues.

xv **The rate of species extinction:** "UN Report: Nature's Dangerous Decline 'Unprecedented'; Species Extinction Rates 'Accelerating,'" Sustainable Development Goals, United Nations, May 6, 2019, https://
www.un.org/sustainabledevelopment/blog/2019/05/nature-decline
-unprecedented-report.

xv **the UN's Intergovernmental Science-Policy Platform:** "UN Report: Nature's Dangerous Decline 'Unprecedented.'"

xvi **The press coverage of that issue:** Hiroko Tabuchi, "In Coronavirus, Industry Sees Chance to Undo Plastic Bag Bans," *New York Times,* March 26, 2020, https://www.nytimes.com/2020/03/26/climate/plastic
-bag-ban-virus.html.

xvi **But there is nothing efficient:** Laura Parker, "Fast Facts About Plastic Pollution," *National Geographic*, December 20, 2018, https://www
.nationalgeographic.com/news/2018/05/plastics-facts-infographics
-ocean-pollution.

xvi **There is nothing efficient:** Brook Larmer, "E-Waste Offers an Economic Opportunity as Well as Toxicity," *New York Times Magazine*, July 5, 2018, https://www.nytimes.com/2018/07/05/magazine/e-waste
-offers-an-economic-opportunity-as-well-as-toxicity.html.

xvi **There is nothing efficient about 40 percent:** Dana Gunders, "Wasted: How America Is Losing Up to 40 Percent of Its Food from Farm to Fork to Landfill," Natural Resources Defense Council, August 16, 2017, https://www.nrdc.org/resources/wasted-how-america-losing-40 -percent-its-food-farm-fork-landfill.

xvii **As Eric Schlosser wrote:** Eric Schlosser, *Fast Food Nation: The Dark Side of the All-American Meal* (New York: Houghton Mifflin, 2001), 260.

xvii **American Sustainable Business Council:** "There Is No Going Back: Creating an Economic System That Works for All," American Sustainable Business Council, October 7, 2020, https://www.asbcouncil .org/post/creating-economic-system-works-alldraft.

xxi **Take the case of Unilever, which:** "Our Sustainable Living Report 2019," Unilever, www.unilever.com/sustainable-living/our-sustainable -living-report-hub.

xxii **They understand that:** Caitlin Mullen, "Millennials Drive Big Growth in Sustainable Products," *Bizwomen*, December 28, 2018. https://www .bizjournals.com/bizwomen/news/latest-news/2018/12/millennials -drive-big-growth-in-sustainable.html.

xxii **In 2018, Unilever's Sustainable Living Brands:** "Unilever's Purpose-Led Brands Outperform," Unilever press release, November 6, 2019, https://www.unilever.com/news/press-releases/2019/unilevers -purpose-led-brands-outperform.html.

xxii **Accenture calculates that the transition:** Peter Lacy and Jakob Rutqvist, *Waste to Wealth: The Circular Economy Advantage* (New York and London: Palgrave Macmillan, 2015), cited in Accenture Strategy's WBCSD "CEO Guide to the Circular Economy," 2015, 4, https://docs.wbcsd.org/2017/06/CEO_Guide_to_CE.pdf.

xxiii **For example, investment bank ING:** "Opportunity and Disruption: How Circular Thinking Could Change US Business Models A Circular Economy Survey," ING 2, https://www.ingwb.com/media/2692501 /ing_us-circular-economy-survey-05-02-2019.pdf.

xxv **As Peter Diamandis and Steven Kotler wrote:** Peter Diamandis and Steven Kotler, *Abundance: The Future Is Better Than You Think* (New York: Simon & Schuster, 2012), xix.

Chapter 1: A Duty to Waste

3 **"When ER nurses are scared":** Pam Deichmann, "Nurses 'Frightened,' Confused by Lack of PPE, Changing Policies," *Iowa Capital Dispatch*, April 14, 2020, https://iowacapitaldispatch.com/2020/04/14 /nurses-frightened-confused-by-lack-of-ppe-changing-policies.

3 **A coalition of doctors:** Zoë Schlanger, "Begging for Thermometers, Body Bags, and Gowns: U.S. Health Care Workers Are Dangerously

Ill-Equipped to Fight COVID-19," *Time*, April 20, 2020, https://time
.com/5823983/coronavirus-ppe-shortage.

4 **An estimated 10.5 percent:** Economic Research Service, U.S. Department of Agriculture, "Food Security in the U.S.: Key Statistics &
Graphics," https://www.ers.usda.gov/topics/food-nutrition-assistance
/food-security-in-the-us/key-statistics-graphics.

4 **"The Capitalist Threat," written by:** George Soros, "The Capitalist
Threat," *Atlantic*, February 1997, https://www.theatlantic.com/maga
zine/archive/1997/02/the-capitalist-threat/376773.

5 **When the home:** *American Bottler* 37 (1917): 42, 47.

6 **95 percent return rate:** Bartow J. Elmore, *Citizen Coke: The Making
of Coca-Cola Capitalism* (New York: W. W. Norton, 2014), 225.

6 **Today, we spend:** Mitch Jacoby, "Why Glass Recycling in the US Is
Broken," *Chemical & Engineering News* 97, no. 6 (February 11, 2019).

7 **Philosopher Marshall McLuhan:** Jib Fowles, *Advertising and Popular Culture* (Thousand Oaks, CA: Sage, 1996), 6.

8 **"Wearing things out":** Earnest Elmo Calkins, "What Consumer Engineering Really Is," in *The Industrial Design Reader*, ed. Carma Gorman
(New York: Allworth Press, 2003), 131.

8 **Calkins explained the strategy:** Calkins, "What Consumer Engineering Really Is," 131.

8 **in 1899:** Susan Strasser, *Waste and Want: A Social History of Trash*
(New York: Picador, 2000), 171.

8 **Shipping them to stores:** "Vintage Ads: Uneeda Biscuit Takes Crackers Out of the Barrel," *Saturday Evening Post*, February 16, 2018,
https://www.saturdayeveningpost.com/2018/02/vintage-advertising
-uneeda-biscuit-takes-crackers-barrel.

9 **However, in 1904:** Nate Barksdale, "How Canned Food Revolutionized
the Way We Eat," History, updated August 22, 2018, https://www.his
tory.com/news/what-it-says-on-the-tin-a-brief-history-of-canned-food.

9 **Fast-forward to today:** Helen Spiegelman and Bill Sheehan, "Unin-
tended Consequences: Municipal Solid Waste Management and the
Throwaway Society," Product Policy Institute, March 2005, 6, doi:
10.13140/RG.2.2.18205.15842.

9 **Disposable rubber gloves:** Walter Brown, "The History of Disposable
Gloves," *AMMEX* (blog), December 12, 2016, https://blog.ammex.com
/the-history-of-disposable-gloves/#.XaEVJEZKiLc.

10 **Exports to Europe rose:** Hugh Rockoff, "U.S. Economy in World War
I," EH.net, n.d., https://eh.net/encyclopedia/u-s-economy-in-world
-war-i.

10 **While the government urged thrift:** University of Washington Li-
braries, entry "Homefront," https://content.lib.washington.edu/exhibits
/WWI/homefront.html.

10 **One was an ad for Nemo corsets:** Trevor Hammond, "Give Up or Give In!: Advertising During World War I," *Fishwrap* (blog), November 13, 2014, https://blog.newspapers.com/give-up-or-give-in-advertising -during-world-war-i.

10 **In the words of advertising guru of the time:** Daniel Pope, "The Advertising Industry and World War I," *Public Historian* 2, no. 3 (Spring 1980): 4–25, https://www.jstor.org/stable/3376987?seq=1#page _scan_tab_contents.

11 **One bulb from the first years:** Jamie Allen, "The Longest Lightbulb," Continent.com, issue 6.1, 2017, 85–86, http://www.continentcontinent .cc/index.php/continent/article/view/271.

11 **the Phoebus cartel:** Markus Krajewski, "The Great Lightbulb Con- spiracy: The Phoebus Cartel Engineered a Shorter-Lived Lightbulb and Gave Birth to Planned Obsolescence," *IEEE Spectrum*, September 24, 2014, https://spectrum.ieee.org/tech-history/dawn-of-electronics /the-great-lightbulb-conspiracy.

12 **Ironically, some of the best evidence:** Vance Packard, *The Waste Makers* (Brooklyn, NY: Ig Publishers, 2011), 70–76.

13 **this "dynamic obsolescence":** Laura Grattan, *Populims's Power: Radical Grassroots Democracy in America* (New York: Oxford Uni- versity Press, 2016), 102.

13 **His motivation came from the overwhelming competition:** Michelle Krebs, "Model A Is a Smashing but Short-lived Success," *Automotive News*, June 16, 2003, https://www.autonews.com/article/20030616/SUB /306160740/model-a-is-a-smashing-but-short-lived-success.

13 **The car was no:** Krebs, "Model A Is a Smashing but Short-lived Suc- cess."

13 **So popular was the Model T:** John Paul, "By the Numbers: Best-selling Vehicles of All Time," Historic Vehicle Association, March 31, 2018, https://www.historicvehicle.org/numbers-best-selling-automobiles -time.

13 **In 1925, Sloan instituted:** David Gartman, *Auto Opium: A Social History of American Automobile Design* (New York: Routledge, 1994).

14 **So devastated was Ford's market share:** Krebs, "Model A Is a Smash- ing but Short-lived Success."

14 **The field of industrial design:** Carroll Gantz, *Refrigeration: A His- tory* (Jefferson, NC: McFarland & Company, 2015), 140.

14 **The goal of their persuasion campaigns:** Gus Lubin, "There's a Stag- gering Conspiracy Behind the Rise of Consumer Culture," *Business Insider*, February 23, 2013, https://www.businessinsider.com/birth-of -consumer-culture-2013-2.

14 **"manipulation of the organized habits":** Edward Bernays, *Propa- ganda* (1928; rept. Brooklyn, NY: Ig Publishing, 2005), 37.

15 **In his treatise "The Engineering of Consent":** Edward Bernays, *The Engineering of Consent* (Norman: University of Oklahoma Press, 1955), 1, https://drwho.virtadpt.net/files/The-Engineering-of-Consent.pdf.

15 **"The group mind":** Bernays, *Propaganda*, 73.

15 **His claims to public:** Amanda Amosa and Margaretha Haglundb, "From Social Taboo to 'Torch of Freedom': The Marketing of Cigarettes to Women," *Tobacco Control* 9, no. 1, https://tobaccocontrol.bmj .com/content/9/1/3.

15 **Another great success:** Lisa Held, "Psychoanalysis Shapes Consumer Culture: Or How Sigmund Freud, His Nephew and a Box of Cigars Forever Changed American Marketing," *Monitor on Psychology* 40, no. 11 (December 2009): 32, https://www.apa.org/monitor/2009/12 /consumer.

15 **For such grand:** Report of the Committee on Recent Economic Changes of the President's Conference on Unemployment, *Recent Economic Changes in the United States*, vols. 1 and 2, National Bureau of Economic Research, 1929, https://www.nber.org/chapters/c4950.pdf.

16 **they were so popular:** Martha L. Olney, "Avoiding Default: The Role of Credit in the Consumption Collapse of 1930," *Quarterly Journal of Economics* 114, no. 1 (February 1999), 321, https://pdfs.semanticscholar .org/885a/5c2adf9eb945e68d0f7cbf477773ec6361f3.pdf9.

16 **Frederick's own magnum opus:** Strasser, *Waste and Want*, 198.

16 **"What is 'progressive obsolescence'":** Christine McGaffey Frederick, *Selling Mrs. Consumer* (New York: The Business Bourse, 1929), 246.

16 **She introduced to the:** Frederick, *Selling Mrs. Consumer*, 79.

16 **There isn't the slightest:** Frederick, *Selling Mrs. Consumer*, 82.

16 **"Mrs. Consumer has billions":** Frederick, *Selling Mrs. Consumer*, 251.

16 **The buildup of excess:** Robert J. Samuelson, "Revisiting the Great Depression," *Wilson Quarterly* (Winter 2012): 36–43.

16 **"If the credit of the United States":** Frederick, *Selling Mrs. Consumer*, 379.

17 **While racing toward the industrial hub:** James Holland, *Normandy '44: D-Day and the Epic 77-Day Battle for France* (New York: Atlantic Monthly Press, 2019), 24–25.

17 **The massive Ford Willow Run facility:** Tim Trainor, "How Ford's Willow Run Assembly Plant Helped Win World War II," *Assembly*, January 3, 2019, https://www.assemblymag.com/articles/94614-how -fords-willow-run-assembly-plant-helped-win-world-war-ii.

18 **with 1.2 million parts:** "Consolidated B-24J Liberator," Collings Foundation, https://www.collingsfoundation.org/aircrafts/consolida ted-b-24-liberator.

18 **Luftwaffe commander General Adolf Galland:** Holland, *Normandy '44*, 37.

18 **Before the war:** Holland, *Normandy '44*, 88.

18 **By war's end:** Thomas D. Morgan, "The Industrial Mobilization of World War II: America Goes to War," *Army History* 30 (Spring 1994): 34, https://www.jstor.org/stable/26304207?seq=4#metadata_info_tab _contents.

18 **Scrap drives exhorting:** Roger Mola, "The Biggest Industrial Boom in U.S. History: How Roosevelt's Fireside Chat Changed American Culture," *Air & Space*, May 2015.

18 **Seventeen million new:** Doris Kearns Goodwin, "The Way We Won: America's Economic Breakthrough During World War II," *The American Prospect*, December 19, 2001, https://prospect.org/health /way-won-america-s-economic-breakthrough-world-war-ii.

19 **Americans had contributed:** Christopher J. Tassava, "The American Economy During World War II," EH.net, https://eh.net/encyclopedia /the-american-economy-during-world-war-ii/.

19 **The campaign caused:** Kathie Smith, "Kitchen of Tomorrow, Viewed Today," *Toledo Blade*, April 6, 2002, https://www.toledoblade.com /opinion/2002/04/07/Kitchen-of-Tomorrow-viewed-today/stories /200204070028.

19 **a Paramount Pictures short film:** "1940s War, Cold War and Consumerism," *Ad Age*, March 28, 2005, https://adage.com/article/75-years -of-ideas/1940s-war-cold-war-consumerism/102702.

19 **While by 1942:** Eric Barnouw, *Tube of Plenty: The Evolution of American Television* (New York: Oxford University Press, 1990), 92.

19 **by 1948 that rose to 2 million:** "1950s: TV and Radio," Encyclopedia .com, https://www.encyclopedia.com/history/culture-magazines/1950s -tv-and-radio.

19 **By 1951, television:** Valerie Bolden-Barrett, "What Caused the Advertising Industry Boom in the 1950s?," *Houston Chronicle*, updated September 04, 2020, https://smallbusiness.chron.com/caused-advertising -industry-boom-1950s-69115.html.

19 ***Newsweek* insightfully predicted:** Stephen J. Phillips, "PLASTIC: MHOF (Monsanto House of the Future)," *Cold War Hothouses: Inventing Postwar Culture, from Cockpit to Playboy*, Beatriz Colomina, Annemarie Brennan, and Jeannie Kim, eds. (Hudson, NY: Princeton Architectural Press, 2004), 97.

20 **As historian Sheldon Garon:** Sheldon Garon, *Beyond Our Means: Why America Spends While the World Saves* (Princeton, NJ: Princeton University Press, 2012), 317–29.

21 **The first popular credit card:** Merrill Fabry, "Now You Know: What Was the First Credit Card?," *Time*, October 19, 2016, https://time.com /4512375/first-credit-card/.

21 ***Life* magazine reported:** "Is Thrift Un-American?," *Life*, May 7, 1956, 40.

21 **Among a wealth of distressing:** Packard, *The Waste Makers*, 32.

21 **Probably the most eloquent:** John Kenneth Galbraith, *The Affluent Society* (Boston: Houghton Mifflin Company, 1998), 116.

21 **Training his ire:** Galbraith, *The Affluent Society*, 124.

22 **a "consumers' republic":** Cohen, "A Consumers' Republic," 236.

22 **In 1934, he wrote a plea:** Scott A. Wolla, "GDP: Does It Measure Up?," *Page One Economics*, Federal Reserve Bank of St. Louis, May 2013, 3.

22 **"the thralldom of a myth":** Galbraith, *The Affluent Society*, 209, 263.

22 **In one of his most stirring:** Robert F. Kennedy, Remarks at the University of Kansas, March 18, 1968, https://www.jfklibrary.org/learn /about-jfk/the-kennedy-family/robert-f-kennedy/robert-f-kennedy -speeches/remarks-at-the-university-of-kansas-march-18-1968.

24 **World Happiness Report:** John Helliwell, Richard Layard, and Jeffrey Sachs, *World Happiness Report* 2019, https://s3.amazonaws.com /happiness-report/2019/WHR19.pdf.

24 **In refuting the dictum:** Alexander Lau and Matthew Green, "Is the Future of the Economy A Doughnut?," OZY, September 19, 2018, https://www.ozy.com/the-new-and-the-next/is-the-future-of-the -economy-a-doughnut/88546.

25 **In the middle is:** Kate Raworth, "A Safe and Just Space for Humanity," Oxfam Discussion Papers, February 2012, https://www-cdn.ox fam.org/s3fs-public/file_attachments/dp-a-safe-and-just-space -for-humanity-130212-en_5.pdf.

26 **"What we are looking at":** Anna Kusmer, "Amsterdam's Coronavirus Recovery Plan Embraces 'Doughnut Economics' for People and the Planet," *The World*, May 11, 2020, https://www.pri.org/stories/2020 -05-11/amsterdam-s-coronavirus-recovery-plan-embraces-doughnut -economics-people-and.

Chapter 2: The Disinformers

27 **"This is just one more":** Sarah Matheson, "NYC Restaurant Owners Tangled in Chemical Lobby," *Epoch Times*, September 13, 2013, updated November 27, 2013, https://www.theepochtimes.com/nyc -restaurant-owners-tangled-in-chemical-lobby_286178.html. Also see "Tell your NYC Council Member: Recycle Foam, Don't Ban It," PS: Re cycleIt.com, August 13, https://www.psrecycleit.com/get_inivolved.

27 **"just another Bloomberg Nanny":** Carl Campanile, "Mayor Bloomberg Wants to Ban Styrofoam," *New York Post*, November 21, 2013, https:// nypost.com/2013/11/21/mayor-bloomberg-wants-to-ban-styrofoam.

27 **Apparently moved by the:** Chester Soria, "Why Ban Foam Food Containers? The Recycling Plan Bloomberg Doesn't Want," *Gotham Gazette*, December 8, 2013, https://www.gothamgazette.com/environ ment/4756-why-ban-foam-containers-the-recycling-plan-bloom berg-doesnt-want.

27 An "industry report": MB Public Affairs, "Fiscal & Economic Impacts of a Ban on Plastic Foam Foodservice and Drink Containers in New York City," March 2013, https://www.plasticfoodservicefacts.com/wp-content/uploads/2017/10/NYC-Foodservice-Impact-Study.pdf.

28 Michael Durant, of the National Federation: Siri Srinivas, "New York Styrofoam Ban Leaves City's Food Carts at Loose Ends," Guardian, January 8, 2015, https://www.theguardian.com/money/us-money-blog/2015/jan/08/new-york-sytrofoam-ban-leaves-food-carts-endangered.

28 "some owners may choose": Sarah Matheson, "Potential Plastic Foam Ban Worries Bronx Restaurant Owners," Epoch Times, August 31, 2013, updated November 27, 2013, https://www.theepochtimes.com/potential-polystyrene-ban-worries-bronx-restaurant-owners_272521.html.

29 In Seattle, for example: Linda D. Nguyen, "An Assessment of Policies on Polystyrene Food Ware Bans" (master's thesis, San Jose State University, Fall 2012), 36, https://scholarworks.sjsu.edu/cgi/viewcontent.cgi?article=1265&context=etd_projects.

29 In California, as: Dustin Gardiner, "Plastic Waste Cuts Probably Headed to California Ballot as Advocates Give Up on Legislature," San Francisco Chronicle, September 12, 2020.

29 The National Resources Defense Council: Eric A. Goldstein, "Foam Containers Disappearing in NYS and Across the Nation," National Resources Defense Council, July 3, 2019, https://www.nrdc.org/experts/eric-goldstein/foam-food-and-beverage-containers-disappearing-nys-and-ac.

29 For example, the Restaurant Action Alliance: See, for example, Recycling Today Staff, "Restaurant Action Alliance of NYC Files Lawsuit Against de Blasio Administration," Recycling Today, September 14, 2017, https://www.recyclingtoday.com/article/restaurant-action-alliance-of-nyc-files-lawsuit-against-de-blasio-administration/; and Claire Lampen, "NYC Ban on Some Single-Use Foam Products Goes into Effect January 1st," Gothamist, November 23, 2018, https://gothamist.com/food/nyc-ban-on-some-single-use-foam-products-goes-into-effect-january-1st.

29 It was an organization: David Giambusso and Brendon Cheney, "Foam Container Industry Spends Big but Loses Support in City Council," Politico, July 27, 2017, https://www.politico.com/states/new-york/city-hall/story/2017/07/26/foam-container-industry-spends-big-but-loses-more-support-in-city-council-113631.

29 When asked to comment: Matheson, "NYC Restaurant Owners Tangled in Chemical Lobby."

30 Another owner who: Matheson, "NYC Restaurant Owners Tangled in Chemical Lobby."

30 **the wife of the Dart CEO:** Giambusso and Cheney, "Foam Contained Industry Spends Big."

30 **including Robert Jackson:** Jeff Coltin, "Robert Jackson Has Been Lobbying for Big Styrofoam," *City & State New York*, July 12, 2018, https://www.cityandstateny.com/articles/politics/news-politics/robert-jackson-has-been-lobbying-big-styrofoam.html.

30 THE RECYCLING PLAN BLOOMBERG DOESN'T WANT: Soria, "Why Ban Foam Food Containers?"

30 DESPITE RECYCLING PUSH: Jessica Lyons Hardcastle, "Despite Recycling Push from Restaurants, Manufacturers, NYC Bans Styrofoam," *Environment + Energy Leader*, January 13, 2015, https://www.environmentalleader.com/2015/01/despite-recycling-push-from-restaurants-manufacturers-nyc-bans-styrofoam.

30 **A piece on the ban:** Jeff Stier, "Don't Ban Styrofoam," *National Review*, July 17, 2018, https://www.nationalreview.com/2018/07/new-york-city-styrofoam-ban-feel-good-measure-costs-more-than-recycling.

30 **On appeal, a subsequent analysis:** *Matter of Restaurant Action Alliance NYC v. City of New York*, October 18, 2018, https://law.justia.com/cases/new-york/appellate-division-first-department/2018/7372-100734-15.html.

31 **The organization really representing:** Vicky Gan, "How the War Against Polystyrene Foam Containers Is Being Won," Route Fifty, January 13, 2015, https://www.routefifty.com/management/2015/01/polystyrene-foam-containers-bans/102678.

31 **If the Dart family:** Karen De Witt, "Exile's Effort to Return Puts Focus on Tax Loophole," *New York Times*, October 1, 1995, https://archive.nytimes.com/query.nytimes.com/gst/fullpage-990CEED81E3EF932A35753C1A963958260.html.

31 **The antiban campaign:** "The Disinformation Playbook," Union of Concerned Scientists, October 10, 2017, updated May 18, 2018, https://www.ucsusa.org/resources/disinformation-playbook.

32 **The term is owed:** Ryan Sager, "Keep Off the Astroturf," *New York Times*, August 18, 2009, https://www.nytimes.com/2009/08/19/opinion/19sager.html.

32 **Patients United Now:** Jane Mayer, "Covert Operations," *New Yorker*, August 23, 2010, https://www.newyorker.com/magazine/2010/08/30/covert-operations.

32 **The group Washington Consumers:** "How Fossil Fuel Lobbyists Used 'Astroturf' Front Groups to Confuse the Public," Union of Concerned Scientists, October 11, 2017, https://www.ucsusa.org/resources/how-fossil-fuel-lobbyists-used-astroturf-front-groups-confuse-public.

32 **In response to Governor Jay Inslee's:** "Washington Consumers for Sound Fuel Policy Issues Statement in Response to Governor Inslee's

Carbon Policy Proposal," Yahoo! Finance, December 17, 2014, https://finance.yahoo.com/news/washington-consumers-sound-fuel-policy-004353066.html.

32 **So profligate was DDT spraying:** Charles C. Mann, "'Silent Spring & Other Writings' Review: The Right and Wrong of Rachel Carson," *Wall Street Journal*, April 26, 2018, https://www.wsj.com/articles/silent-spring-other-writings-review-the-right-and-wrong-of-rachel-carson-1524777762.

33 **For that important public service:** Linda Lear, introduction to Rachel Carson, *Silent Spring* (New York: Mariner Books, 2002), xvii.

33 **Monsanto distributed a brochure:** Mark Stoll, "Industrial and Agricultural Interests Fight Back," Rachel Carson's *Silent Spring* Virtual Exhibition, Environment and Society Portal, updated 2020, http://www.environmentandsociety.org/exhibitions/rachel-carsons-silent-spring/industrial-and-agricultural-interests-fight-back.

33 **Rather, she commended:** Carson, *Silent Spring*, 275, 184.

33 **His review was titled:** Bryan Walsh, "How Silent Spring Became the First Shot in the War Over the Environment," *Time*, September 25, 2012, http://science.time.com/2012/09/25/how-silent-spring-became-the-first-shot-in-the-war-over-the-environment.

33 **Again and again Carson:** GM Watch, "Rachel Carson Centennial: The Continuing Campaign by Monsanto & the Agri-Toxics Lobby to Silent Dissent," Organic Consumers Association, May 27, 2007, https://www.organicconsumers.org/news/rachel-carson-centennial-continuing-campaign-monsanto-agri-toxics-lobby-silence-dissent.

33 **A former secretary of agriculture:** Stoll, Rachel Carson's *Silent Spring* Virtual Exhibition, "The Personal Attacks on Rachel Carson as a Woman Scientist."

33 **The general counsel:** Eliza Griswold, "How 'Silent Spring' Ignited the Environmental Movement," *New York Times Magazine*, September 21, 2012, https://www.nytimes.com/2012/09/23/magazine/how-silent-spring-ignited-the-environmental-movement.html.

34 **In truth, Carson championed democratization:** Carson, *Silent Spring*, 127.

34 **in response to "obvious evidence":** Carson, *Silent Spring*, 13.

34 **In an hour-long:** Sarah Grey, "In Defense of Rachel Carson," *International Socialist Review* 57 (January–February 2008), www.isreview.org/issues/57/feat-rachelcarson.shtml.

34 **In the next decade:** Mann, "'Silent Spring & Other Writings' Review."

36 **"There's so many lies":** Soria, "Why Ban Foam Food Containers?"

36 **Alan Shaw of Plastics Recycling:** Colin Staub, "NYC Sued over Foam Ban . . . Again," *Plastics Recycling Update*, September 13, 2017, https://resource-recycling.com/plastics/2017/09/13/nyc-sued-foam-ban.

36 **We see the tactic:** Donald Brown, "Ethical Analysis of Disinformation Campaign's Tactics: (1) Think Tanks (2) PR Campaigns (3) Astroturf Groups, and (4) Cyber-Bullying Attacks," Ethics and Climate, February 10, 2012, https://ethicsandclimate.org/2012/02/10/ethical _analysis_of_disinformation_campaigns_tactics_1_think_tanks_2 _pr_campaigns_3_astroturf_groups.

36 **Take the fevered complaint:** Oversight Hearings on the U.S. Postal Service—1994: Hearings Before the Committee on Post Office and Civil Service, House of Representatives, One Hundred Third Congress, First Session, April 14, 19, 26; May 17, 24; June 8, 1994.

37 **an estimated 5.6 million tons:** Elisabeth Leamy, "How to Stop Junk Mail and Save Trees—and Your Sanity," *Washington Post*, February 13, 2018, https://www.washingtonpost.com/lifestyle/home/how-to-stop -junk-mail-and-save-trees—and-your-sanity/2018/02/12/6000e4c4 -05d9-11e8-b48c-b07fea957bd5_story.html.

37 **"certain outstanding entomologists":** Carson, *Silent Spring*, 259.

38 **According to internal company memos:** J. Cook et al., "America Misled: How the Fossil Fuel Industry Deliberately Misled Americans About Climate Change," George Mason University Center for Climate Change Communication, 2019, https://www.climatechangecom munication.org/america-misled.

38 **A 1988 internal memo lays out the strategy:** J. Cook et al., "America Misled," 7.

38 **The Union of Concerned Scientists:** "The Disinformation Playbook."

38 **As for Exxon:** J. Cook et al., "America Misled," 7.

40 **Farmers have received subsidies:** Dan Charles, "Farmers Got Billions from Taxpayers in 2019, and Hardly Anyone Objected," *All Things Considered*, NPR, December 31, 2019, https://www.npr.org /sections/thesalt/2019/12/31/790261705/farmers-got-billions-from -taxpayers-in-2019-and-hardly-anyone-objected.

40 **Executives began contemplating:** All details in this account of Exxon's entrance and exit from solar power are from Andrea Hsu, "How Big Oil of the Past Helped Launch the Solar Industry of Today," *All Things Considered*, NPR, September 30, 2019, https://www.npr.org /2019/09/30/763844598/how-big-oil-of-the-past-helped-launch-the -solar-industry-of-today.

41 **As Naomi Klein highlighted:** Naomi Klein, *This Changes Everything: Capitalism vs. the Climate* (New York: Simon & Schuster, 2014), 145.

41 **boom in the methane-leaking process:** Heather Richards, "Is U.S. Shale Facing an 'Unmitigated Disaster'?," *E&E News*, September 19, 2019, https://www.eenews.net/stories/1061136849.

41 **Smaller oil and gas producers:** Clifford Krauss, "U.S. Oil Companies Find Energy Independence Isn't So Profitable," *New York Times*,

June 30, 2019, https://www.nytimes.com/2019/06/30/business/energy -environment/oil-companies-profit.html.

41 **As for the giants:** Jordan Blum, "Exxon Mobil Profits Plunge with Lower Oil, Fuels, Plastics Prices," *Houston Chronicle*, April 26, 2019, https://www.houstonchronicle.com/business/energy/article/Exxon -Mobil-profits-plunge-with-lower-oil-price-13797725.php.

41 **its share price has stayed virtually flat:** Krauss, "U.S. Oil Companies Find Energy Independence Isn't So Profitable."

41 **Estimates of the total annual subsidies:** David Coady et al., "Global Fossil Fuel Subsidies Remain Large: An Update Based on Country-Level Estimates," IMF Working Papers, International Money Fund, May 2, 2019, https://www.imf.org/en/Publications/WP/Issues/2019 /05/02/Global-Fossil-Fuel-Subsidies-Remain-Large-An-Update-Based -on-Country-Level-Estimates-46509.

41 **In *Dark Money*:** Jane Mayer, *Dark Money: The Hidden History of the Billionaires Behind the Rise of the Radical Right* (New York: Anchor Books, 2017), 124.

42 **In market tests:** Rebecca Altman, "American Beauties," *Topic*, no. 14, "Earth, Wind & Fire," August 2018, https://www.topic.com/american -beauties.

42 **Many shoppers outright hated:** Alexis Petru, "A Brief History of the Plastic Bag," *Triple Pundit*, November 5, 2014, https://www.triplepun dit.com/story/2014/brief-history-plastic-bag/39406.

42 **One press release:** Jube Shiver Jr., "Supermarket Dilemma: Battle of the Bags: Paper or Plastic?," *Los Angeles Times*, June 22, 1986.

42 **So intent was Exxon:** See, for example, patent application available at https://patents.google.com/patent/US4840335.

43 **the media has forecast its demise:** "USA: Recycling Generates 117 Billion US-Dollar per Year," Global Recycling, https://global-recycling .info/archives/2172.

43 **Throughout 2018 and 2019:** Headlines quoted are from Chaz Miller, "Who Killed Recycling?," Waste360, June 1, 2018, https://www.waste360 .com/recycling/who-killed-recycling; and "Why America's Recycling Is in the Dumps," CBS News, October 10, 2018, https://www.cbsnews .com/news/why-americas-recycling-industry-is-in-the-dumps.

44 **the second largest waste collector:** Henry Grabar, "Recycling Isn't About the Planet. It's About Profit," *Slate*, April 5, 2019, https://slate .com/business/2019/04/recycling-dead-planet-profit-americans -commodities-china.html.

44 ***The New York Times* reported:** Michael Corkery, "As Costs Skyrocket, More U.S. Cities Stop Recycling," *New York Times*, March 16, 2019, https://www.nytimes.com/2019/03/16/business/local-recycling -costs.html.

44 **"mountains of paper":** "Why America's Recycling Is in the Dumps."

44 **Yet, when a spokesperson:** Colin Staub, "Investments Contradict 'End of Recycling' Headlines," *Resource Recycling*, March 26, 2019, https://resource-recycling.com/recycling/2019/03/26/investments -contradict-end-of-recycling-headlines.

44 **A comprehensive effort:** Cole Rosengren et al., "How Recycling Has Changed in All 50 states," WasteDive, updated November 15, 2019, https://www.wastedive.com/news/what-chinese-import-policies -mean-for-all-50-states/510751.

45 **The host of doomsaying articles:** Mary Esch, "China's Ban on Scrap Imports a Boon to US Recycling Plants," Associated Press, May 18, 2019, https://apnews.com/3d3d86a09b9647b795295e77355877db.

45 **In an article published by:** Edward Humes, "The U.S. Recycling System Is Garbage," *Sierra*, June 26, 2019, https://www.sierraclub.org/si erra/2019-4-july-august/feature/us-recycling-system-garbage.

46 **Probably none has done so more:** John Tierney, "Recycling Is Garbage," *New York Times Magazine*, June 30, 1996, https://www.nytimes .com/1996/06/30/magazine/recycling-is-garbage.html.

47 **In one case the hole:** Paul Heyne, *"Are Economists Basically Immoral?" and Other Essays on Economics, Ethics, and Religion*, eds. A. M. C. Waterman and Geoffrey Brennan (Indianapolis, IN: Liberty Fund, 2008). See chapter 26, "Recycling and Dumping: A Case Study," https://oll.libertyfund.org/titles/heyne-are-economists-basically -immoral-and-other-essays-on-economics-ethics-and-religion /simple.

47 **in another "15 square":** Jerry Taylor, "Recycling Is Not the Answer," *Roll Call*, February 24, 1992, https://www.cato.org/publications/com mentary/recycling-is-not-answer.

47 **contrarian provocateur John Stossel:** "John Stossel vs Lauren Singer Analysis," Don't Be a Tree Stump, October 22, 2016, updated February 20, 2017, http://www.dontbeatreestump.com/2016/10/22/john -stossel-vs-lauren-singer-analysis.

47 **Professor Clark Wiseman:** A. Clark Wiseman, "U.S. Waste Paper and Recycling Policies: Issues and Effects, Resources for the Future," Washington, D.C.: Energy and Natural Resources Division, Resources for the Future, 1990, 2.

47 **Not specified was that:** Daniel K. Benjamin, "The Benefits of Climate Change," *PERC Reports* 25, no. 3 (Fall 2007).

47 **Journalist Chris Mooney:** Chris Mooney, "Libertarian Rhapsody," *The American Prospect*, December 19, 2001, https://prospect.org/fea tures/libertarian-rhapsody.

47 **if he were "an equal-opportunity":** Mooney, "Libertarian Rhapsody."

48 **One wonders whether Tierney's:** Dr. Richard A. Denison and John F. Ruston, "Debunking the Myths of the Anti-Recyclers," EDF Letter,

EDR.org, July 18, 1996. The reprise article by John Tierney is "The Reign of Recycling," *New York Times*, October 3, 2015, https://www.nytimes.com/2015/10/04/opinion/sunday/the-reign-of-recycling.html.

49   **The forces advocating circularity:** Paul Hawken, *The Ecology of Commerce* (New York: HarperCollins, 1993), 15.

Chapter 3: Circularity Innovators Forge Ahead

51   **Inflation soared, with prices:** Orlando Letelier, "The 'Chicago Boys' in Chile: Economic Freedom's Awful Toll," *The Nation*, first published August 1976, reprinted September 21, 2016, https://www.the nation.com/article/archive/the-chicago-boys-in-chile-economic -freedoms-awful-toll.

53   **That was done by:** "The Circular Economy: Interview with Walter Stahel," *MakingIt*, July 5, 2013, https://www.makingitmagazine.net /?p=6793.

53   **Stahel came up with:** The account of Stahel's contribution is based on numerous sources, including: Thibaut Wautelet, "The Concept of Circular Economy: Its Origins and Its Evolution" (working paper, January 2018), https://www.researchgate.net/publication/322555840 _The_Concept_of_Circular_Economy_its_Origins_and_its_Evolu tion; Jose-Luis Cardoso, "The Circular Economy: Historical Grounds," in *Changing Societies: Legacies and Challenges*, vol. 3, *The Diverse Worlds of Sustainability*, eds. Ana Delicado et al. (Lisbon, Portugal: Instituto de Ciências Sociais, 2018); and Ellen MacArthur Foundation, *Towards the Circular Economy: Economic and Business Rationale for an Accelerated Transition*, 2013, https://www.ellenmacarthurfounda tion.org/assets/downloads/publications/Ellen-MacArthur-Founda tion-Towards-the-Circular-Economy-vol.1.pdf.

54   **the Apollo 8 Saturn V rocket:** Apollo 8, "Part 2: Apollo 8—In the Beginning There Was Liftoff," NASA.gov, December 19, 2018, https://www.nasa.gov/feature/goddard/2018/apollo-8-in-the-beginning -there-was-liftoff.

54   **the most powerful spacecraft constructed:** Robin McKie, "The Mis sion That Changed Everything," *Guardian*, November 29, 2008, https:// www.theguardian.com/science/2008/nov/30/apollo-8-mission.

54   **The craft would need to travel:** Robert Kurson, *Rocket Men: The Daring Odyssey of Apollo 8 and the Astronauts Who Made Man's First Journey to the Moon* (New York: Random House, 2019), 49.

54   **It would then need to modulate:** Kurson, *Rocket Men*, 226.

54   **If the craft's single engine:** Kurson, *Rocket Men*, 49.

54   **So dangerous was the mission:** Kurson, *Rocket Men*, 129.

55   **"Oh my god":** "Transcripts of Earthrise: The 45th Anniversary," NASA.gov, https://svs.gsfc.nasa.gov/vis/a000000/a004100/a004129 /G2013-102_Earthrise_MASTER_youtube_hqTranscripts.html.

55 **Frank Borman recalled thinking:** Larry Getlen, "How Apollo 8 Captured One of the Most Famous Photos in History," *New York Post*, May 16, 2017, https://nypost.com/2017/05/16/how-apollo-8-captured-one-of-the-most-famous-photos-in-history.

55 **As Robert Kurson wrote:** Kurson, *Rocket Men*, 308.

55 **this new "spaceman economy":** Kenneth Boulding, "The Economics of the Coming Spaceship Earth," in ed. Henry Jarrett, *Environmental Quality in a Growing Economy* (Baltimore: Resources for the Future/ Johns Hopkins University Press, 1966), 3–14.

56 **a "steady-state economy":** Herman E. Daly, *Beyond Growth: The Economics of Sustainable Development* (Boston: Beacon Press, 1996), 31.

56 **Reading *Silent Spring*:** Lissa Harris, "The Economic Heresy of Herman Daly," *Grist*, April 10, 2003, https://grist.org/article/bank.

57 **He persuasively described:** Barry Commoner, *The Closing Circle: Nature, Man and Technology* (Mineola, NY: Dover Publications, 2020), 36.

58 **The stately sentinel:** Bess Lovejoy, "15 of the World's Most Famous Trees," Mental Floss, May 14, 2016, updated July 23, 2019, https://www.mentalfloss.com/article/63667/15-worlds-most-famous-trees.

58 **the tree has withstood:** Geoffrey Migiro, "What and Where Is Major Oak?," World Atlas, May 10, 2018, https://www.worldatlas.com/articles/what-and-where-is-major-oak.html.

58 **Consider the feat pulled off:** Henning Grann, "The Industrial Symbiosis at Kalundborg, Denmark," in *The Industrial Green Game: Implications for Environmental Design and Management*, ed. Deanna J. Richards (Washington, D.C.: National Academy Press, 1997), 117–23, https://www.nap.edu/read/4982/chapter/10.

59 **A 2015 analysis:** World Bank Group, *Enhancing China's Regulatory Framework for Eco-Industrial Parks: Comparative Analysis of Chinese and International Green Standards*, April 2019, http://documents.worldbank.org/curated/en/950911554814522228/pdf/Enhancing-China-s-Regulatory-Framework-for-Eco-Industrial-Parks-Comparative-Analysis-of-Chinese-and-International-Green-Standards.pdf.

59 **Impressive progress in scaling:** "Eco-Industrial Parks Emerge as an Effective Approach to Sustainable Growth," The World Bank, January 23, 2018, https://www.worldbank.org/en/news/feature/2018/01/23/eco-industrial-parks-emerge-as-an-effective-approach-to-sustainable-growth.

60 **One beautiful exemplar:** Edwina Langley, "Could London Take Inspiration from Rotterdam's 'Zero-Waste' BlueCity?" *Evening Standard*, October 15, 2019, https://www.standard.co.uk/futurelondon/the plasticfreeproject/could-london-take-inspiration-from-rotterdams-zerowaste-bluecity-a4261306.html.

60 **She bemoaned the evisceration:** Carson, *Silent Spring*, 66.

62 **The concept of natural capital:** E. F. Schumacher, *Small Is Beautiful: Economics as if People Mattered* (New York: Harper Perennial, 2010), 15.

62 **His opening line:** Schumacher, *Small Is Beautiful*, 13.

62 **The term "circular economy":** Stephen Smith, "David Pearce: Environmental Economist Whose Market-Based Ideas Caught the Changing Tide of the 1980s," *Guardian*, September 21, 2005, https://www.theguardian.com/science/2005/sep/22/highereducation.guardianobituaries.

62 **Pearce, with coauthor:** David Pearce and R. Kerry Turner, *Economics of Natural Resources and the Environment* (Baltimore, MD: Johns Hopkins University Press, 1990), 29.

63 **"How is it that":** Paul Hawken, *The Ecology of Commerce* (New York: HarperCollins, 1993), 13.

63 **"It turns out":** Amory B. Lovins, L. Hunter Lovins, and Paul Hawken, "A Road Map for Natural Capitalism," *Harvard Business Review*, July-August 2007, https://hbr.org/2007/07/a-road-map-for-natural-capitalism.

64 **"ultimate ecosystem engineers":** Ben Goldfarb, "Beavers Are the Ultimate Ecosystem Engineers," *Sierra*, July 3, 2018, https://www.sierraclub.org/sierra/2018-4-july-august/feature/beavers-are-ultimate-ecosystem-engineers.

64 **So keen are:** Julia Zorthian,"The True History Behind Idaho's Parachuting Beavers," *Time*, October 23, 2015, https://time.com/4084997/parachuting-beavers-history.

64 **In England, where:** Claire Marshall, "Beaver Families Win Legal 'Right to Remain,'" BBC, August 6, 2020, https://www.bbc.com/news/science-environment-53658375.

64 **One analysis concludes:** *Towards the Circular Economy*, 21.

65 **One calculation: the value:** David Pearce, "The Economic Value of Forest Ecosystems," *Ecosystem Health* 7, no. 4 (December 2001): 291.

65 **one recent estimate:** Simon G. Potts et al., *The Assessment Report on Pollinators, Pollination and Food Production*, Intergovernmental Science-Policy Platform on Biodiversity and Ecosystem Services, 2017, xxx, https://ipbes.net/sites/default/files/downloads/pdf/2017_pollination_full_report_book_v12_pages.pdf.

66 **"That's when I knew":** Jeffrey Hollender, "Fixing the Freest Marketplace Money Can Buy," *Stanford Social Innovation Review*, March 30, 2016, https://ssir.org/articles/entry/fixing_the_freest_marketplace_money_can_buy.

66 **According to one analysis:** Sarah Axe and Danielle Nierenberg, "The True Cost of Food: An Excerpt from *Nourished Planet*," Resilience, June 12, 2018, https://www.resilience.org/stories/2018-06-12/the-true-cost-of-food-an-excerpt-from-nourished-planet.

66   **One corporate leader:** Richard Benson, "Jochen Zeitz Saved Puma. Now He's Trying to Fix Global Business," *Wired*, November 5, 2019, https://www.wired.co.uk/article/jochen-zeitz.

67   **No mere proselytizer:** Walter R. Stahel, "The Circular Economy," *Nature*, March 23, 2016, https://www.nature.com/news/the-circular -economy-1.19594.

68   **Rolls-Royce, a leader:** "'Power by the Hour': Can Paying Only for Performance Redefine How Products Are Sold and Serviced?," Knowledge@Wharton, February 21, 2017, https://knowledge.wharton.upenn .edu/article/power-by-the-hour-can-paying-only-for-performance -redefine-how-products-are-sold-and-serviced.

68   **Tire maker Michelin:** Chloe Renault, Frederic Dalsace, and Wolfgang Ulaga, "Michelin Fleet Solutions: From Selling Tires to Selling Kilometers," Case Centre, 2010, revised November 29, 2012, https:// www.thecasecentre.org/educators/products/view?id=96546.

68   **A particularly gratifying:** "Selling Light as a Service," Ellen MacArthur Foundation, https://www.ellenmacarthurfoundation.org/case-studies /selling-light-as-a-service.

68   **Amsterdam's Schiphol Airport:** Mark Faithfull, "Pay-as-You-Go Lighting Arrives at Amsterdam's Schiphol Airport," LuxReview, April 20, 2015, https://www.luxreview.com/2015/04/20/pay-as-you-go -lighting-arrives-at-amsterdam-s-schiphol-airport. For the amount of electricity use by the airport, see "Reducing Energy Consumption," Royal Schipol Group, https://www.schiphol.nl/en/schiphol-group/page /reducing-energy-consumption.

69   **Though a remanufactured:** Peter Lacy and Jakob Rutqvist, *Waste to Wealth: The Circular Economy Advantage* (New York and London: Palgrave Macmillan, 2015), 9.

69   **What a problem crying out:** Vladimir Puskas, "ToothPASTe: Let's Redesign It," OpenIdeo, updated July 26, 2017, https://challenges .openideo.com/challenge/circular-design/ideas/toothpaste-1.

69   **Thankfully, Colgate has stepped up:** Jordan Valinsky, "Colgate's New Recyclable Toothpaste Tube Is Nearly Ready. It Took 5 Years to Develop," *CNN Business*, June 19, 2019, https://www.cnn.com/2019 /06/19/business/colgate-recyclable-tube-trnd/index.html.

70   **When it comes to reducing waste:** Nadine von Moltke-Todd, "How This Environmentally Conscious Entrepreneur Is Following Her Passions," Entrepreneur.com South Africa, August 1, 2019, https:// www.entrepreneur.com/article/337504.

70   **A runaway success:** Jeff Beer, "How Patagonia Grows Every Time It Amplifies Its Social Mission," *Fast Company*, February 21, 2018, https://www.fastcompany.com/40525452/how-patagonia-grows -every-time-it-amplifies-its-social-mission.

70 **An inspiring case:** "Built Positive Principles Play Key Role in Venlo City Hall's Sustainable Design Mission," Cradle to Cradle Productions Innovation Institute, January 28, 2017, https://www.c2ccertified.org/news/article/built-positive-principles-play-key-role-in-venlo-city-halls-sustainable-des.

71 **Given that the demolition:** "Sustainable Management of Construction and Demolition Materials," U.S. Environmental Protection Agency, https://www.epa.gov/smm/sustainable-management-construction-and-demolition-materials.

71 **Regarding designing for renewal:** Irina Vinnitskaya, "VanDusen Botanical Garden Visitor Centre / Perkins+Will," *ArchDaily*, March 20, 2012, https://www.archdaily.com/215855/vandusen-botanical-garden-visitor-centre-perkinswill.

72 **"The wind was up to 30 knots":** Ellen MacArthur, *Taking on the World: A Sailor's Extraordinary Solo Race Around the World* (New York: McGraw Hill, 2003), 279.

73 **"When you set off":** "Navigating the Circular Economy: A Conversation with Dame Ellen MacArthur," McKinsey & Company, February 1, 2014, https://www.mckinsey.com/business-functions/sustainability/our-insights/navigating-the-circular-economy-a-conversation-with-dame-ellen-macarthur.

73 **"As I finish this message":** MacArthur, *Taking on the World*, 286.

Chapter 4: For the Love of Forests

78 **Its success has led:** Pratt Industries, "Overview & Growth" in *Sustainable Packaging & Display Solutions*, http://www.prattindustries.com/downloads/Pratt-Industries-Brochure.pdf.

78 **Paper mushed and reconstituted:** "Pratt Industries the Works," video, YouTube, May 15, 2011, https://www.youtube.com/watch?v=qXABlDk4cYU.

78 **But he and his wife:** Account of Leon Pratt's story summarized from James Kirby and Rod Myer, *Richard Pratt: One Out of the Box: The Secrets of An Australian Billionaire* (Melbourne: John Wiley & Sons, 2009), chapter 1.

79 **He's always ready:** Chase Peterson-Withorn, "Recycling Riches: How Australian Billionaire Anthony Pratt Is Getting Wealthier Off Americans' Trash," *Forbes*, July 29, 2015, https://www.forbes.com/sites/chasewithorn/2015/07/29/recycling-riches-how-australian-billionaire-anthony-pratt-is-getting-wealthier-off-americans-trash/#13e172965988.

84 **Luellen introduced his wonder:** Subba Yerra, "A Brief History of . . . The Disposable Paper Cup," Medium, August 17, 2019, https://medium.com/@subba.ry/a-brief-history-of-the-disposable-paper-cup-8976a657025e.

84   **To force their hands:** Journal of the Senate of Texas of the Regular Session of the 32d Legislature, Austin, January 10, 1911, 270.

85   **As Colleen Chapman:** Clare Goldsberry, "Starbucks Launches 'Moonshot for Sustainability,'" *Plastics Today*, March 23, 2018, https://www.plasticstoday.com/packaging/starbucks-launches-moonshot-sustainability.

87   **When Hearst bought:** W. Joseph Campbell, *Yellow Journalism: Puncturing the Myths, Defining the Legacies* (Westport, CT: Praeger Publishers, 2001), 6.

87   **The volume of newspapers:** William A. Dill, "Growth of Newspapers in the United States" (master's thesis, University of Oregon, 1908), 12, https://kuscholarworks.ku.edu/bitstream/handle/1808/21361/dill_1928_3425151.pdf?sequence=1.

87   **In 1975, *Bloomberg BusinessWeek*:** "The Office of the Future," *Bloomberg BusinessWeek*, June 30, 1975.

87   **While it's not in the least:** Abigail J. Sellen and Richard H. R. Harper, *The Myth of the Paperless Office* (Cambridge, MA: MIT Press, 2001), 13.

87   **A 1999 report estimated:** "Computers Have Not Caused a Reduction in Paper Usage or Printing," HistoryofInformation.com, 1999, https://www.historyofinformation.com/detail.php?entryid=1394.

87   **annual worldwide paper consumption:** Kazimierz Przybysz et al., "Application of Poplar Pulps as Cost-Effective Addition to Production of Paper," *International Journal of Microstructure and Materials Properties* 8 (June 2014): 584–91.

87   **Overall paper use:** "Paper, Paper, Paper, Paper, Paper, Paper, Paper, PAPER . . . TOO MUCH PAPER!," *Think Green News*, Summer 2016, https://www.upstate.edu/green/pdf/tg-news/tg-news-summer-2016.pdf.

87   **Many of us really like:** Jill Barshay, "Evidence Increases for Reading on Paper Instead of Screens," *The Hechinger Report*, August 12, 2019, https://hechingerreport.org/evidence-increases-for-reading-on-paper-instead-of-screens.

87   **The long-anticipated decline:** David J. Unger, "American Reams: Why A 'Paperless World' Still Hasn't Happened," *Guardian*, December 29, 2017, https://www.theguardian.com/news/2017/dec/29/american-reams-why-the-paperless-world-hasnt-happened.

88   **That's a difficult number:** "Paper Recycling Facts," Recycling at USI, University of Southern Indiana, https://www.usi.edu/recycle/paper-recycling-facts.

88   **The boom in e-commerce:** Smithers, "Five Key Trends That Are Changing the Future of the Corrugated Packaging Market," Smithers.com, n.d., https://www.smithers.com/resources/2019/jan/trends-changing-the-corrugated-packaging-market.

88 **The good news:** Author interview with Gary Scott, February 18, 2020.

89 **At one plant:** Account based on "Paper Recycling: Market Deinked Pulp: A Tour of a Paper Recycling Facility with Dr. Richard Venditti," video, YouTube, July 12, 2017, https://www.youtube.com/watch?v= ZnE4wSUnjhs&t=1103s.

90 **The Resolute plant:** "ND Paper LLC Completes Acquisition of Resolute Forest Products Recycled Pulp Mill in Fairmont, West Virginia," *Cision PR Newswire*, November 1, 2018, https://www.prnewswire .com/news-releases/nd-paper-llc-completes-acquisition-of-resolute -forest-products-recycled-pulp-mill-in-fairmont-west-virginia -300742159.html.

91 **Yes, Canada's British Columbia is home:** "The Rainforest Coast Region," BC's Great Wild Spaces, https://www.spacesfornature.org /greatspaces/rainforest.html.

91 **Meanwhile, forests in the southeastern U.S.:** Fen Montaigne, "Why Keeping Mature Forests Intact Is Key to the Climate Fight," Yale-Environment360, Yale School of the Environment, October 15, 2019, https://e360.yale.edu/features/why-keeping-mature-forests-intact -is-key-to-the-climate-fight.

91 **"Trees unite to":** Peter Wohlleben, *The Hidden Life of Trees* (Vancouver, BC, and Berkeley, CA: Greystone Books, 2016), xi.

91 **Researchers have found that beech trees:** Wohlleben, *The Hidden Life of Trees*, 15.

92 **Forest ecologist Suzanne Simard discovered:** Diane Toomey, "Exploring How and Why Trees 'Talk' to Each Other," YaleEnvironment360, Yale School of the Environment, September 1, 2016, https://e360.yale .edu/features/exploring_how_and_why_trees_talk_to_each_other.

92 **They are said to account for about 50 percent:** Rhett A. Butler, "Rainforest Information," Mongabay, August 14, 2020, https://rainforests .mongabay.com.

92 **One estimate of the potential:** Adam Aton, "Diverse Forests Capture More Carbon," E&E News, *Scientific American*, October 5, 2018, https://www.scientificamerican.com/article/diverse-forests-capture -more-carbon.

92 **To this aim:** See https://www.weforum.org/agenda/2020/01/one-tril lion-trees-world-economic-forum-launches-plan-to-help-nature-and -the-climate/#:~:text=The%20World%20Economic%20Forum%20 has%20launched%20a%20global,businesses%20and%20individuals %20in%20a%20%22mass-scale%20nature%20restoration%

93 **Mark Ashton of the Yale School:** Author interview with Mark Ashton, February 18, 2020.

93 **His father was forest botany legend:** Alvin Powell, "Peter Ashton: A Legacy Written in Trunk, Limb and Leaf," *Harvard Gazette*, July 17, 2008, https://news.harvard.edu/gazette/story/2008/07/peter-ashton-a-legacy-written-in-trunk-limb-and-leaf.

94 **In territory in northern Australia:** Thomas Fuller, "Reducing Fire, and Cutting Carbon Emissions, the Aboriginal Way," *New York Times*, January 16, 2020, https://www.nytimes.com/2020/01/16/world/australia/aboriginal-fire-management.html.

95 **So prodigious is:** "9 Rainforest Facts Everyone Should Know," Rainforest Alliance, June 20, 2019, https://www.rainforest-alliance.org/pictures/9-rainforest-facts-everyone-should-know.

95 **Forests also help prevent flooding:** Wohlleben, *The Hidden Life of Trees*, 245.

95 **An analysis of one program:** Stefano Pagiola, "Paying for Ecosystem Services, a Successful Approach to Reducing Deforestation in Mexico," *World Bank Blogs*, March 4, 2019, https://blogs.worldbank.org/latinamerica/paying-ecosystem-services-successful-approach-reducing-deforestation-mexico.

95 **In California, a "tropical forest":** Daniel Nepstad, "How to Help Brazilian Farmers Save the Amazon," *New York Times*, December 24, 2019.

96 **Consider that after colonization:** "Capacity of North American Forests to Sequester Carbon," ScienceDaily.com, July 13, 2018, https://www.sciencedaily.com/releases/2018/07/180713093531.htm.

96 **Carolina Levis, one of the researchers:** Robinson Meyer, "The Amazon Rainforest Was Profoundly Changed by Ancient Humans," *Atlantic*, March 7, 2017, https://www.theatlantic.com/science/archive/2017/03/its-now-clear-that-ancient-humans-helped-enrich-the-amazon/518439.

96 **The Ceibo Alliance is a nonprofit:** "Alianza Ceibo/The Ceibo Alliance," Namati.org, https://namati.org/network/organization/alianza-ceibo-the-ceibo-alliance.

96 **Cofounder Nemonte Nenquimo:** Nemonte Nenquimo, "This Is My Message to the Western World—Your Civilisation Is Killing Life on Earth," *Guardian*, October 12, 2020.

97 **Many indigenous forest:** "9 Rainforest Facts Everyone Should Know."

97 **As Suzanne Simard says:** Suzanne Simard, "How Trees Talk to Each Other," TED Talk, TEDSummit June 2016.

Chapter 5: Greener Grocery

99 **He had planned:** Cornell College of Agricultural and Life Sciences, "Unparalleled Integrity, Conviction Earn Excellence In IPM Award for Legendary 'Rat Czar,'" press release, February 7, 2017, https://

nysipm.cornell.edu/about/we-give-awards/2016-excellence-ipm
-award-winners.

99 **As a graduate student:** David Segal, "New York Tackles Its Gnawing
Rat Problem, Purdue Alum BS '77, MS '80, PhD '95," Purdue University, March 20, 2007, https://www.entm.purdue.edu/news/corrigan2
.html.

100 **Teaching clients his "Sherlock Holmes approach":** Brad Harbison,
"Bobby Corrigan Discusses the 'Sherlock Holmes Approach' to Rodent IPM," Pest Control Technology, January 8, 2020, https://www
.pctonline.com/article/corrigan-sherlock-holmes-rodent-ipm.

101 **an estimated 40 percent:** Dana Gunders et al., *Wasted: How America
Is Losing up to 40 Percent of Its Food from Farm to Fork to Landfill,*
(Natural Resources Defense Council, August 2017), 4, https://www.nrdc
.org/sites/default/files/wasted-2017-report.pdf.

101 *Consumer Reports* **estimates:** Christina Troitino, "Americans Waste
About a Pound of Food a Day, USDA Study Finds," *Forbes*, April 23,
2018, https://www.forbes.com/sites/christinatroitino/2018/04/23/ameri
cans-waste-about-a-pound-of-food-a-day-usda-study-finds/#1da
7e664ec3b.

101 **a pound of food:** Gunders et al., *Wasted*, 4.

101 **households are said:** Gunders et al., *Wasted*, 10.

101 **In fact, food waste:** Chris Peak, "6 High-Tech Innovations That
Could Solve Our Food-Waste Woes," Nation Swell, January 16, 2017,
https://nationswell.com/tech-solves-food-waste-woes.

101 **Paul Hawken's Project Drawdown:** Amelia Nierenberg, "One Thing
Your City Can Do: Reduce Food Waste," *New York Times*, December 11,
2019, https://www.nytimes.com/2019/12/11/climate/nyt-climate-news
letter-food-waste.html.

101 **What they are is:** Rachel Jackson, "Most People Waste More Food
Than They Think—Here's How to Fix It," *National Geographic*, April
24, 2019, https://www.nationalgeographic.com/environment/2019/04
/people-waste-more-food-than-they-think-psychology.

102 **The general belief:** Jackson, "Most People Waste More Food Than
They Think."

102 **Overstocking is so common:** Gunders et al., *Wasted*, 22.

102 **former president of:** Harrison Jacobs, "Why Grocery Stores Like
Trader Joe's Throw Out So Much Perfectly Good Food," *Business Insider*, October 15, 2014, https://www.businessinsider.com/why-grocery
-stores-throw-out-so-much-food-2014-10.

102 **Stores suffer considerable:** Barbara Bean-Mellinger, "What Is the
Profit Margin for a Supermarket?," *Houston Chronicle*, updated November 14, 2018, https://smallbusiness.chron.com/profit-margin-super
market-22467.html.

103   **Yet, when Stop and Shop:** Gunders et al., *Wasted*, 24.

105   **They are said to have made quick work:** John Lanchester, "The Case Against Civilization," *New Yorker*, September 11, 2017, https://www.newyorker.com/magazine/2017/09/18/the-case-against-civilization.

105   **Their diet, for the most part:** Daniel J. DeNoon, "7 Rules for Eating," WebMD, March 23, 2009, https://www.webmd.com/food-recipes/news/20090323/7-rules-for-eating#1.

105   **the transition to cultivation:** S. Katz, F. Maytag, and M. Civil, (1991). "Brewing an Ancient Beer," *Archaeology*, 44(4), 24–33. Retrieved January 21, 2021, from http://www.jstor.org/stable/41765984.

106   ***Forbes* reported in May 2020:** Nicole F. Roberts, "Number of Food Insecure Households More Than Doubles as Food Banks Struggle," *Forbes*, May 26, 2020.

106   **Yet it's estimated:** Alissa Link, "Food and Tech: Solutions to Recover and Redistribute Food Waste," Hunter College New York City Food Policy Center, June 26, 2019, https://www.nycfoodpolicy.org/food-and-tech-solutions-to-recover-redistribute-food-waste.

106   **An estimated $1.3 billion:** Link, "Food and Tech."

107   **The USDA's National School Lunch Program:** Jonathan Bloom, "Waste Not, Want Not," *Grist*, November 28, 2018, https://grist.org/article/schools-waste-5-million-a-day-in-uneaten-food-heres-how-oakland-is-reinventing-the-cafeteria.

107   **One solution that requires nothing:** Bloom, "Waste Not, Want Not."

107   **One of these:** Katerina Bozhinova, "16 Apps Helping Companies and Consumers Prevent Food Waste," GreenBiz, October 12, 2018, https://www.greenbiz.com/article/16-apps-helping-companies-and-consumers-prevent-food-waste.

107   **Israeli company Evigence:** Tom Karst, "Evigence Sensors Seeking Quality, Food Safety Uses in Fresh Produce," ThePacker.com, January 29, 2020, https://www.thepacker.com/news/food-safety-markets/marketing/evigence-sensors-seeking-quality-food-safety-uses-fresh-produce.

108   **Researchers at Imperial College:** Caroline Brogan, "Food Freshness Sensors Could Replace 'Use-By' Dates to Cut Food Waste," Imperial College London, June 5, 2019, http://www.imperial.ac.uk/news/191413/food-freshness-sensors-could-replace-use-by.

109   **Day after day:** Allison Aubrey, "Landfill of Lettuce: Why Were These Greens Tossed Before Their Time?," New Hampshire Public Radio, June 16, 2015, https://www.nhpr.org/post/landfill-lettuce-what-happens-salad-past-its-prime#stream/0.

110   **Apeel Sciences, which produces:** Justin Jaffe, "Apeel's Longer-Lasting Fruit Means Fewer Trips to the Grocery Store," CNet, May

26, 2020, https://www.cnet.com/features/apeels-longer-lasting-fruit
-means-fewer-trips-to-the-grocery-store.

111 **A friend of Albert's:** E. Fairlie Watson, "The Lessons of the East,"
Organic Gardening Magazine 13, no. 8 (1948), http://journeytoforever
.org/farm_library/howard_memorial.html#Watson.

112 **In one of those ironic coincidences:** Michael Pollan, *The Omnivore's
Dilemma* (New York: Penguin Press, 2007), 41.

112 **In 1944, Norman Borlaug:** Gregg Easterbrook, "Forgotten Benefactor
of Humanity," *Atlantic*, January 1997, https://www.theatlantic.com
/magazine/archive/1997/01/forgotten-benefactor-of-humanity/306101.

112 **the population soared:** Mark Kurlansky, *The Food of a Younger Land*
(New York: Riverhead Books, 2010), introduction.

113 **How densely packed?:** Joseph P. Harner III and James P. Murphy,
"Planning Cattle Feedlots," Kansas State University, Department of
Biological and Agricultural Engineering, https://bookstore.ksre.ksu
.edu/pubs/MF2316.pdf; see also, "Balancing Your Animals with Your
Forage," Natural Resources Conservation Service, U.S. Department
of Agriculture, https://www.nrcs.usda.gov/Internet/FSE_DOCUME
NTS/stelprdb1167344.pdf.

113 **What's more, because:** Barry Estabrook, "Feedlots vs. Pastures: Two
Very Different Ways to Fatten Beef Cattle," *Atlantic*, December 28,
2011, https://www.theatlantic.com/health/archive/2011/12/feedlots-vs
-pastures-two-very-different ways-to-fatten-beef-cattle/250543.

113 **Today, only 27 percent:** Mark Hyman, *Food Fix* (New York: Hachette,
2020), 23.

113 **At the beginning of the twentieth:** Mark Wilson, "Infographic: In 80
Years, We Lost 93% of Variety in Our Food Seeds," *Fast Company*,
May 11, 2012.

114 **In a bizarre testament to how:** Ward Sinclair, "Under Missouri: A
Monument to the Output of the American Cow," *Washington Post*,
December 21, 1981, https://www.washingtonpost.com/archive/politics
/1981/12/21/under-missouri-a-monument-to-the-output-of-the
-american-cow/7d5376f4-d53e-4209-b6cb-99cd833612eb.

114 **just 1 percent:** Catherine Greene et al., "Growing Organic Demand
Provides High-Value Opportunities for Many Types of Producers,"
Amber Waves, USDA Economic Research Service, February 6, 2017,
https://www.ers.usda.gov/amber-waves/2017/januaryfebruary/grow
ing-organic-demand-provides-high-value-opportunities-for-many
-types-of-producers.

114 **Coverage of the industrialization:** Kimberly Amadeo, "Farm Subsidies with Pros, Cons, and Impact," TheBalance.com, updated June
29, 2020, https://www.thebalance.com/farm-subsidies-4173885.

114 **From 1995 to 2017:** Amadeo, "Farm Subsidies with Pros, Cons, and Impact."

114 **For family farms:** George Jared, "CDC: 'Farm Stress,' Suicides a Rising Rural Health Concern," TB&P, May 14, 2019, https://talkbusiness .net/2019/05/cdc-farm-stress-suicides-a-rising-rural-health-concern.

115 **The number of farmers:** Ryan Stockwell, "The Family Farm in the Post-World War II Era," (PhD diss., University of Missouri-Columbia, 2008), 2.

115 **A shocking recent study:** Maria Godoy, "It's Not Just Salt, Sugar, Fat," *All Things Considered*, NPR, May 16, https://www.npr.org/sections /thesalt/2019/05/16/723693839/its-not-just-salt-sugar-fat-study-finds -ultra-processed-foods-drive-weight-gain.

115 **The science of food flavoring:** Eric Schlosser, *Fast Food Nation: The Dark Side of the All-American Meal* (New York: Houghton Mifflin, 2001), 125–27.

116 **As Michael Moss relates:** Michael Moss, *Salt Sugar Fat* (New York: Random House, 2013), 49.

116 **In a revelation that staggers the mind:** Moss, *Salt Sugar Fat*, xxvi, 10, 321.

116 **Mark Hyman writes that, to get out of this bind:** Hyman, *Food Fix*, 7.

116 **I reached Gabe:** Account based on interview with the author, April 20, 2020.

117 **founded the Rodale Institute:** Rodale Institute, "Our Story," https:// rodaleinstitute.org/about/our-story.

119 **As Mark Hyman highlighted:** Hyman, *Food Fix*, 46. The study is "General Mills: Accounting for Soil Impacts in Carbon Footprints," Quantis, Case Studies, https://quantis-intl.com/casestudy/general -mills/.

119 **Now, in most fields:** "Gabe Brown in Idaho 2: Keys to Building a Healthy Soil," video, YouTube, March 2, 2015, https://www.youtube .com/watch?v=K51DaRKET20.

120 **His yield results:** Rodale Institute, "The Farming Systems Trial Celebrating 30 Years," https://rodaleinstitute.org/wp-content/uploads/fst -30-year-report.pdf.

120 **Investment firm Steward Partners:** Lisa Held, "With $100, You Too Can Invest in Regenerative Agriculture," *Civil Eats*, October 2, 2019, https://civileats.com/2019/10/02/with-100-you-too-can-invest -in-regenerative-agriculture/.

121 **As of 2019, $47.5 billion:** Christi Electris et al., *Soil Wealth: Investing in Regenerative Agriculture Across Asset Classes*, (Croatan Institute, July 2019), http://www.croataninstitute.org/images/publications/soil -wealth-2019.pdf.

121 **Overseas, the UN teamed:** Chris Prentice, "Rabobank, U.N. Launch $1 Billion Fund to Boost Sustainable Farming," Reuters, October

16, 2017, https://www.reuters.com/article/us-rabobank-sustainability /rabobank-u-n-launch-1-billion-fund-to-boost-sustainable-farming -idUSKBN1CL2Y2.

121 **Dairy products giant Danone:** Heather Clancy, "General Mills, Danone Dig Deeper into Regenerative Agriculture with Incentives, Funding," *GreenBiz*, February 20, 2020, https://www.greenbiz.com /article/general-mills-danone-dig-deeper-regenerative-agriculture -incentives-funding.

121 **Yair Teller, one of the founders:** Account based on the author's multiple conversations with Yair Teller.

123 **New York City has been running:** Dan Sandoval, "New York Region to Host Large Anaerobic Digester," *Recycling Today*, March 22, 2019, https://www.recyclingtoday.com/article/anerobic-digester-food -waste-long-island-brookhaven/.

124 **In Philadelphia:** Andrew Maykuth, "Philly Refiner Plans $120M Plant to Convert Food Scraps to Fuel for Trucks and Buses," *Philadelphia Inquirer*, August 29, 2018, https://www.inquirer.com/philly/business /energy/philadelphia-energy-solutions-food-waste-digester-methane -gas-fuel-20180828.html#:~:text=now%20Resend%20email-,Philly%20 refiner%20plans%20%24120M%20plant%20to%20convert%20food, fuel%20for%20trucks%20and%20buses&text=RNG%20Energy%20 Solutions%2C%20which%20built,food%20waste%20into%20transpor tation%20fuel.

124 **North of Los Angeles:** LJ Dawson, "How Cities Are Turning Food Into Fuel," *Politico*, November 21, 2019.

124 **Sir Albert Howard wrote:** Sir Albert Howard, *The Soil and Health* (Lexington: University of Kentucky Press, 2006), 257.

Chapter 6: The Sustainable Closet

125 **When I went to:** Account based on interview with the author, September 3, 2020.

126 **She was determined:** Nicole Bassett, "The Future of Business Is Circular," video, TEDxBend, August 27, 2019, https://www.youtube.com /watch?v=rUZfIBJHzvY.

127 **"To make a fashion company":** Author interview with Caroline Brown, May 13, 2020.

127 **In recent memory:** Michael Safi and Dominic Rushe, "Rana Plaza, Five Years On: Safety of Workers Hangs in Balance in Bangladesh," *Guardian*, April 24, 2018, https://www.theguardian.com/global-deve lopment/2018/apr/24/bangladeshi-police-target-garment-workers -union-rana-plaza-five-years-on.

128 **As fashion writer Dana Thomas reports:** Dana Thomas, *Fashionopolis: The Price of Fast Fashion and the Future of Clothes* (New York: Penguin Press, 2019), 41.

128 **As for environmental havoc:** "UN Helps Fashion Industry Shift to Low Carbon," UN Climate Change, September 6, 2018, https://unfccc.int/news/un-helps-fashion-industry-shift-to-low-carbon.

128 **Add to that the 17 to 20 percent:** Pamela Ravasio, "How Can We Stop Water from Becoming a Fashion Victim?," *Guardian*, March 7, 2012, https://www.theguardian.com/sustainable-business/water-scarcity-fashion-industry.

128 **So much water:** Western Bonime, "20 Fashion Revolutionaries Making Sustainability a Fashion Must-Have Luxury," *Forbes*, June 10, 2018.

128 **With so many of our clothes now made from:** Julien Boucher and Damien Friot, *Primary Microplastics in the Oceans*, (International Union for Conservation of Nature, 2017), https://www.iucn.org/content/primary-microplastics-oceans.

128 **The estimate is:** Laura Paddison, "Single Clothes Wash May Release 700,000 Microplastic Fibres, Study Finds," *Guardian*, September 26, 2016, https://www.theguardian.com/science/2016/sep/27/washing-clothes-releases-water-polluting-fibres-study-finds.

129 **It's also especially vulnerable:** "A New Textiles Economy: Redesigning Fashion's Future," Ellen MacArthur Foundation, 2017, 38, http://www.ellenmacarthurfoundation.org/publications.

129 **The production of one cotton shirt:** Deborah Drew and Genevieve Yehounme, "The Apparel Industry's Environmental Impact in 6 Graphics," World Resources Institute, July 5, 2017, https://www.wri.org/blog/2017/07/apparel-industrys-environmental-impact-6-graphics.

129 **Uzbekistan has become:** Glynis Sweeny, "Fast Fashion Is the Second Dirtiest Industry in the World, Next to Big Oil," *EcoWatch*, August 17, 2015, https://www.ecowatch.com/fast-fashion-is-the-second-dirtiest-industry-in-the-world-next-to-big—1882083445.html.

129 **Shocked by the apocalyptic:** Brian Merchant, "World's 4th Largest Lake Is Now 90% Dried Up," *TreeHugger*, updated October 11, 2018, https://www.treehugger.com/worlds-th-largest-lake-is-now-dried-up-pics-video-4858036.

129 **Further contributing to:** "#WearNext—Make Fashion Circular Joins Forces with City of New York and Fashion Industry to Tackle Clothing Waste," Ellen MacArthur Foundation, March 4, 2019, https://www.ellenmacarthurfoundation.org/news/wearnext-make-fashion-circular-joins-forces-with-city-of-new-york-and-fashion-industry-to-tackle-clothing-waste.

129 **the annual haul is calculated:** Veronique Greenwood, "Keeping Clothes Out of the Garbage," *Anthropocene*, https://www.anthropocenemagazine.org/sustainablefashion.

129 **Meanwhile, about 20 percent:** Haley Smith Recer, "The Cost of Dead Inventory: Retail's Dirty Little Secret," *Business of Fashion*, July 11,

2017, https://www.businessoffashion.com/articles/opinion/the-cost-of -dead-inventory-retails-dirty-little-secret.

129 **Oxfam reports that:** Elizabeth L. Cline, *The Conscious Closet: The Revolutionary Guide to Looking Good While Doing Good* (New York: Plume, 2019), Kindle, 30.

130 **The Ellen MacArthur Foundation:** Ellen MacArthur Foundation, *A New Textiles Economy: Redesigning Fashion's Future*, 2017, 21, http:// www.ellenmacarthurfoundation.org/publications.

130 **Harvard chemistry professor:** "Wallace Carothers and the Development of Nylon," American Chemical Society, https://www.acs.org /content/acs/en/education/whatischemistry/landmarks/carother spolymers.html.

130 **DuPont also spared:** Keri Blakinger, "A Look Back at Some of the Coolest Attractions at the 1939 World's Fair," *New York Daily News*, April 30, 2016, https://www.nydailynews.com/new-york/queens/back -attractions-1939-world-fair-article-1.2619155.

130 **While some press coverage:** "Japanese Fear Curtailed Silk Exports to U.S.: Extensive Use of Nylon in Hosiery Trade Is Seen Ahead," *St. Cloud Times*, April 13, 1939.

131 **Advertised as "so durable":** "Nylon Hosiery Put On Sale in Wilmington," *Reading Times*, October 25, 1939.

131 **The original name:** Audra J. Wolfe, "Nylon: A Revolution in Textiles," Science History Institute, October 2, 2008, https://www.sciencehistory .org/distillations/nylon-a-revolution-in-textiles.

131 **One woman reportedly:** Robert M. Andrews, "The 50-Year Run of Nylon Stockings: 'Nobody Said, "Eureka 3/8,"'"Associated Press, January 16, 1988, https://apnews.com/article/2a1caf94afa0d947920bfb007 a26edc6.

131 **The resultant stocking:** William Cooper, "Nylon Mob, 40,000 Strong, Shrieks and Sways for Mile," *Pittsburgh Press*, June 13, 1946.

131 **As author Susannah Handley:** Susannah Handley, *Nylon: The Story of a Fashion Revolution* (Baltimore, MD: Johns Hopkins University Press, 2000), 40.

131 **The environmental implications:** "Miss 1939 Proudly Wears Dust Her Mother Despised," *Wilmington Morning News*, October 25, 1939.

132 **The most popular:** Kaity Cornellier, "What Is Polyester? The 8 Most Vital Questions Answered," Contrado, September 25, 2019, https:// www.contrado.com/blog/what-is-polyester.

132 **The Scott Paper company:** Natalie McKane, "Here Are the 1960s Fashion Trends We'd Like to Nominate for a Comeback," Messy Nessy, February 6, 2019, https://www.messynessychic.com/2019/02 /06/paper-dresses-pvc/; Livia Caligor, "Paper Fashion in the 1960s: The Genesis of Fast Fashion," *Cornell Fashion Textile Collection* (blog),

March 17, 2018, https://blogs.cornell.edu/cornellcostume/2018/03/17/paper-fashion-in-the-1960s-the-genesis-of-fast-fashion.

133 **Soon paper pantsuits:** Ena Naunton, "It's an Easy Snip from Pantsuit to Bikini," *Miami Herald*, October 16, 1966.

133 **But by 1968:** Nancy Hayfield, "First Time, I Was Sure Dress Would Fall Apart," *Delaware County Daily Times*, June 25, 1966.

133 **The Consumer Price Index:** Mark J. Perry, "Chart of the Day: The CPI for Clothing Has Fallen by 3.3% over the Last 20 Years, While Overall Prices Increased by 63.5%," *Carpe Diem* (blog), American Enterprise Institute, October 12, 2013, https://www.aei.org/carpe-diem/chart-of-the-day-the-cpi-for-clothing-has-fallen-by-3-3-over-the-last-20-years-while-overall-prices-increased-by-63-5.

133 **A midrange Brooks Brothers:** Eric Wilson, "Dress for Less and Less," *New York Times*, May 29, 2008, https://www.nytimes.com/2008/05/29/fashion/29PRICE.html.

133 **One result is that:** Keila Tyner, "The Case for Fewer—but Better—Clothes," Quartz, March 31, 2014, https://qz.com/189904/the-case-for-fewer-but-better-clothes/.

134 **That's despite the fact:** Drew and Yehounme, "The Apparel Industry's Environmental Impact in 6 Graphics."

134 **Meanwhile, the average:** Elizabeth L. Cline, *Overdressed: The Shockingly High Cost of Cheap Fashion* (New York: Penguin/Portfolio, 2012), 121.

134 **Psychological studies have:** Karen Pine, "30 Fascinating Facts About Fashion Psychology," *HuffPost*, updated October 5, 2014, https://www.huffingtonpost.co.uk/karen-pine/fashion-psychology_b_5650424.html.

134 **One study, for example:** Nicolas Guéguen and Céline Jacob, "Clothing Color and Tipping," *Journal of Hospitality and Tourism* 38, no. 2 (2012): 275–80.

136 **A 2020 Boston Consulting Group Report:** Fashion for Good and Boston Consulting Group, "Financing the Transformation in the Fashion Industry," 2020, 25, https://fashionforgood.com/wp-content/uploads/2020/01/FinancingTheTransformation_Report_FINAL_Digital-1.pdf.

136 **In 1993, the year Zeitz:** Fredrick Obura, "Ex-PUMA Boss Appointed to Kenya Wildlife Service Board," *The Standard*, August 1, 2018, https://www.standardmedia.co.ke/business/article/2001290294/former-puma-boss-lands-job-at-kws.

136 **Zeitz recalls that:** Jochen Zeitz and Anselm Grün, *The Manager and the Monk: A Discourse on Prayer, Profit, and Principles* (New York: Jossey Bass, 2013), 33.

136 **At an estimated market size:** "The Athleisure Market Size Was Valued at $155.2 Billion in 2018 and Is Expected to Reach $257.1 Billion by 2026, Registering a CAGR of 6.7% from 2019 to 2026," Cision PR Newswire, December 2, 2019, https://www.prnewswire.com/news-releases/the -athleisure-market-size-was-valued-at-155-2-billion-in-2018-and-is -expected-to-reach-257-1-billion-by-2026—registering-a-cagr-of -6-7-from-2019-to-2026–300967335.html.

137 **He commissioned Puma's:** Author interview with Jochen Zeitz, February 13, 2020.

137 **Puma's share price:** Richard Benson, "Jochen Zeitz Saved Puma. Now He's Trying to Fix Global Business," *Wired*, November 5, 2019, https://www.wired.co.uk/article/jochen-zeitz.

137 **The group's EP&L:** "Kering Reports on Sustainability Progress and Shows Very Promising First Results Towards Meeting 2025 Targets," Kering, January 30, 2020, https://www.kering.com/en/news/kering -reports-on-sustainability-progress-and-shows-very-promising -first-results-towards-meeting-2025-targets.

138 **An especially appealing transparency innovation:** Sarah Spellings, "The Brand Making It Easier Than Ever to Know Where Your Clothes Come From," *The Cut*, January 30, 2020, https://www.thecut.com /2020/01/the-brand-making-it-easy-to-see-where-your-clothes -come-from.html.

138 **Start-up Bolt Threads:** Bia Bezamat, "Bolt Threads and Stella McCarthy Introduce Mushroom Leather Handbag," *Current Daily*, https:// thecurrentdaily.com/2018/04/17/bolt-threads-stella-mccartney -mushroom-leather/.

138 **Piñatex is another:** Eleanor Lawrie, "The Bizarre Fabrics That Fashion Is Betting On," BBC News, September 9, 2019, https://www.bbc .com/news/business-49550263.

139 **Soybean cashmere, made from:** "Soy Clothing Superior Softness Feels Like Your Second Skin," Cool Organic Clothing, https://www .cool-organic-clothing.com/soy-clothing.html.

139 **Banana fiber textiles:** Avneet Kaur, "Banana Fibre: A Revolution in Textiles," Fibre2Fashion.com, November 2015, https://www.fibre 2fashion.com/industry-article/7654/banana-fibre-a-revolution -in-textiles.

139 **Daughter and father team:** Author meetings and interviews with Renana and Oded Krebs.

140 **Renana was told:** Rosie Manins, "Designer Hoping Moss Creations Grow on Judges," *Otago Daily Times*, March 26, 2012, https://www .odt.co.nz/news/dunedin/designer-hoping-moss-creations-grow -judges.

141 **The Negev is a world-leading cultivator:** "Global Algae Product Market Is Expected to Reach USD 5.38 Billion by 2025: Fior Markets," Globe Newswire.com, January 29, 2020, https://www.globenewswire.com /news-release/2020/01/29/1976747/0/en/Global-Algae-Product-Market -is-Expected-to-Reach-USD-5-38-Billion-by-2025-Fior-Markets.html.

142 **Eileen Fisher is sourcing wool:** Steff Yotka, "'The Biggest Thing We Can Do Is Reduce'—Eileen Fisher Shares a Vision for a Sustainable Future," *Vogue*, April 22, 2020, https://www.vogue.com/article/eileen -fisher-amy-hall-sustainabiity-horizon-2030.

143 **She got the idea:** "Meet Shilla Kim-Parker of Thrilling," *VoyageLA*, January 8, 2019, http://voyagela.com/interview/meet-shilla-kim-parker -of-thrilling.

143 **That's vital, because as Worn Again's:** Nina Notman, "Recycling Clothing the Chemical Way," ChemistryWorld, January 27, 2020, https://www.chemistryworld.com/features/recycling-clothing-the -chemical-way/4010988.article.

143 **H&M has become:** Brian Taylor, "HKRITA Rolls Up Sleeves to Tackle Textile Recycling," *Recycling Today*, June 26, 2019, https:// www.recyclingtoday.com/article/hkrita-textile-recycling-hong -kong-g2g-handm/.

144 **One of the most impressive:** Author interview with Kristy Caylor, June 1, 2020.

145 **Well, consider that millennial households:** Allan Joseph, "How Kristy Caylor's 'For Days' Could Reboot the Fashion Industry's OS," Futur404, June 5, 2018.

146 **The potential for the model:** Jeremy Sporn and Stephanie Tuttle, "5 Surprising Findings About How People Actually Buy Clothes and Shoes," *Harvard Business Review*, June 6, 2018, https://hbr.org/2018 /06/5-surprising-findings-about-how-people-actually-buy-clothes -and-shoes.

Chapter 7: I've Got One Word for You, Benjamin

148 **One of those intrepid adventurers:** Daniel Fastenberg, "Trash Found Littering Ocean Floor in Deepest-Ever Sub Dive," Reuters, May 13, 2019, https://www.reuters.com/article/us-environment-pollu tion/trash-found-littering-ocean-floor-in-deepest-ever-sub-dive -idUSKCN1SJ241.

148 **Shortly thereafter, research:** Sarah Gibbens, "Plastic Proliferates at the Bottom of the World's Deepest Ocean Trench," *National Geographic*, May 13, 2019.

149 **While we might:** "The Great Pacific Garbage Patch Is Not What You Think It Is," *The Swim*, video, Seeker YouTube Channel, December 3, 2018, https://www.youtube.com/watch?v=6HBtl4sHTqU.

150 **Research we conducted:** "Accelerating Circular Supply Chains for Plastics," Closed Loop Partners, April 2019, https://www.closedloop partners.com/wp-content/uploads/2020/01/CLP_Circular_Supply _Chains_for_Plastics.pdf.

150 **There was nothing inevitable:** Jeffrey L. Meikle, *American Plastic: A Cultural History* (New Brunswick, NJ: Rutgers University Press, 1997), xiii.

150 **There are naturally forming plastics:** "History and Future of Plastics," Science History Institute, https://www.sciencehistory.org/the -history-and-future-of-plastics.

151 **As Susan Freinkel recounts:** Susan Freinkel, *Plastic: A Toxic Love Story* (New York: Houghton Mifflin, 2011), 13.

151 **Its scarcity even:** Lin Poyer, "The Ngatik Massacre: Documentary and Oral Traditional Accounts," *The Journal of Pacific History* 20, no. 1 (1985): 4–22, doi: 10.1080/00223348508572502.

152 **The government massively:** Freinkel, *Plastic*, 25.

152 **The group launched:** "What's Behind the Boom in Plastics?," *National Post* (Toronto), January 29, 1944, 13.

152 **SPI held the:** Clare Goldsberry, "NPE's origin story, promoting and defending plastics," *Plastics Today*, March 1, 2018, https://www.plas ticstoday.com/injection-molding/npes-origin-story-promoting-and -defending-plastics.

152 **A newspaper reporter summed up the bounty:** .E. Magnell, "New Products and New Methods," *Hartford Courant*, April 29, 1946.

152 **The show's organizer, Ronald Kinnear:** "Host of New Uses in Plastics Shown," *New York Times*, April 23, 1946.

152 **So beloved was:** Freinkel, *Plastic*, 8.

153 **A Scripps-Howard article:** *Knoxville Sentinel*, July 19, 1947, 4.

153 **One result was:** Peter G. Ryan, "A Brief History of Marine Litter Research," in Bergmann M., Gutow L., Klages M. eds., *Marine Anthropogenic Litter*, (Springer International Publishers, 2015), https://doi.org /10.1007/978-3-319-16510-3_1 https://link.springer.com/chapter/10.1007 /978-3-319-16510-3_1.

153 **As reported by:** Root, "Inside the Long War."

154 **Jeffrey Meikle, who extolled:** Root, "Inside the Long War."

154 **As revealed in:** *Frontline*, episode 14, "Plastic Wars," March 31, 2020, PBS, https://www.pbs.org/wgbh/frontline/film/plastic-wars/tran script/.

155 **Starved for stock:** Joe Sandler Clarke and Emma Howard, "US Plastic Waste Is Causing Environmental Problems at Home and Abroad," Unearthed, Greenpeace, May 10, 2018, https://unearthed .greenpeace.org/2018/10/05/plastic-waste-china-ban-united-states -america/.

157 **The commercialization of biodegradables:** Closed Loop Partners, "Accelerating Circular Supply Chains for Plastics."

158 **This is a perfect:** Tim Dickinson, "Planet Plastic: How Big Oil and Big Soda Kept a Global Environmental Calamity a Secret for Decades," *Rolling Stone*, March 3, 2020, https://www.rollingstone.com/culture/culture-features/plastic-problem-recycling-myth-big-oil-950957/.

159 **The founders of New York State–based:** Lana Chehabeddine, "Fungi-Inspired Companies Could Play a New Role in Sustainability," *Green-Biz*, April 8, 2020, https://www.greenbiz.com/article/fungi-inspired-companies-could-play-new-role-sustainability.

160 **As *The New York Times*:** Steven Kurutz, "Life Without Plastic Is Possible. It's Just Very Hard," *New York Times*, February 16, 2019, https://www.nytimes.com/2019/02/16/style/plastic-free-living.html.

160 **Speaking of beer, both Carlsberg:** Lauren Brown West-Rosenthal, "22 Big Companies That Are Getting Rid of Plastic for Good," *Reader's Digest*, May 24, 2020, https://www.rd.com/article/companies-getting-rid-plastic.

160 **Coca-Cola invested $25 million:** "Reuse: A Closer Look at Coca-Cola Brazil's Unique Returnable Bottle Initiative," Packaging Europe, February 11, 2020, http://packagingeurope.com/coca-cola-brazil-returnable-bottle-initiative.

161 **An innovative alternative:** Joel Makower, "Loop's Launch Brings Reusable Packaging to the World's Biggest Brands," *GreenBiz*, January 24, 2019; Kevin J. Ryan, "The Containers for Your Most Basic Household Products Are About to Look a Lot Different, Thanks to This Company," *Inc.*, March 27, 2019; Karine Vann, "Loop's Quest for Reuse Dominance Has Only Gotten More Ambitious During the Pandemic," *WasteDive*, August 19, 2020.

161 **The high value:** Ian C. Freestone, "The Recycling and Reuse of Roman Glass: Analytical Approaches," *Journal of Glass Studies* 57 (2015): 29–40.

162 **The typical glass bottle:** "Glass vs Plastic: The Facts," Trayak, http://trayak.com/glass-vs-plastic-the-facts.

162 **So preservative is glass:** Brad Smithfield, "The 1,650-year-old Speyer Wine Is the Oldest Bottle of Wine in the World," *Vintage News*, February 16, 2016, https://www.thevintagenews.com/2016/02/16/the-1650-year-old-speyer-wine-is-the-oldest-bottle-of-wine-in-the-world-2.

162 **With the resurgence:** Scott Breen and Jay Siegel, "How a Unique Industry Collaboration Is Bottling a New Future for U.S. Glass Recycling," *GreenBiz*, October 12, 2018, https://www.greenbiz.com/article/how-unique-industry-collaboration-bottling-new-future-us-glass-recycling.

162 **I was fascinated to:** Author interview with Joel Schoening, June 29, 2020.

163 **The bottle is now:** Author interview with Joel Gunderson, July 1, 2020.

164 **"The New Plastics Economy":** "The New Plastics Economy: Rethinking the Future of Plastics," Ellen MacArthur Foundation, World Economic Forum, and McKinsey & Company, January 19, 2016, 17.

Chapter 8: Gold Mines in Our Hands

165 **On April 11, 2019:** "SpaceX Falcon Heavy, All 3 Cores Landed! Successful Launch of Arabsat 6A," video, YouTube, April 11, 2019, https://www.youtube.com/watch?v=-dCln8noi5g.

166 **By refurbishing and relaunching:** Brad Tuttle, "Here's How Much It Costs for Elon Musk to Launch a SpaceX Rocket," *Money*, February 6, 2018, http://money.com/money/5135565/elon-musk-falcon-heavy-rocket-launch-cost.

166 **Also included in the payload:** "LightSail," The Planetary Society, https://www.planetary.org/sci-tech/lightsail.

166 **As NASA administrator:** "SpaceX's Reusable Falcon 9 Rocket: Space Launch LIVE," video, YouTube, May 29, 2020, https://www.youtube.com/watch?v=YzmoSIoHsY.

166 **All of which is:** Michael Sheetz and Amanda Macius, "Elon Musk: SpaceX Is Chasing the 'Holy Grail' of Completely Reusing a Rocket," CNBC, November 5, 2019, https://www.cnbc.com/2019/11/05/elon-musk-completely-reusing-rockets-is-spacexs-holy-grail.html.

167 **Every year, an:** All statistics from "A New Circular Vision for Electronics: Time for a Global Reboot," World Economic Forum, January 2019, 5, http://www3.weforum.org/docs/WEF_A_New_Circular_Vision_for_Electronics.pdf.

167 **Federico Magalini, a leading expert:** Angela Chen, "Why Failing to Recycle Electronics Leaves Gold Mines Untapped," The Verge, July 3, 2018, https://www.theverge.com/2018/4/23/17270960/electronic-waste-urban-mining-materials-recycling.

167 **The Mponeng Gold Mine:** Matthew Hart, "A Journey into the World's Deepest Gold Mine," *Wall Street Journal*, December 13, 2013, https://www.wsj.com/articles/a-journey-into-the-world8217s-deepest-gold-mine-1386951413.

168 **John Shegerian's entrepreneurial:** Multiple meetings and author interview with John Sheregian, June 5, 2020.

171 **As a graduate fellow:** Multiple meetings and author interview with Matanya Horowitz, June 6, 2020.

172 **Back in 1995:** Nicholas Negroponte, *Being Digital* (New York: Knopf, 1995), 2.

172 **Amazon plans to:** Michael Sheetz, "Amazon Will Invest over $10 Billion in Its Satellite Internet Network after Receiving FCC Authorization," CNBC, July 30, 2020, https://www.cnbc.com/2020/07/30/fcc-authorizes-amazon-to-build-kuiper-satellite-internet-network.html.

172 **As for robots:** Negroponte, *Being Digital*, 204.

172 **Starwood Hotels has:** Jordan Crook, "Starwood Introduces Robotic Butlers at Aloft Hotel in Cupertino," *TechCrunch*, August 13, 2014, https://techcrunch.com/2014/08/13/starwood-introduces-robotic-butlers-at-aloft-hotel-in-palo-alto.

172 **Boston Dynamics has:** Erico Guizzo, "How Boston Dynamics Is Redefining Robot Agility," *IEEE Spectrum*, November 27, 2019, https://spectrum.ieee.org/robotics/humanoids/how-boston-dynamics-is-redefining-robot-agility.

173 **As Dr. Kai-Fu Lee:** Kai-Fu Lee, *AI Superpowers: China, Silicon Valley, and the New World Order* (New York: Houghton Mifflin, 2018), 11, 13.

173 **My favorite story:** "Research Report: The Next Product You Design Might Be a Service Thanks to the IOT," Engineering.com, https://www.engineering.com/ResourceMain.aspx?resid=860, 9–10.

174 **The Spanish passenger railroad:** "Research Report: The Next Product You Design Might Be a Service Thanks to the IOT," 11–12.

174 **One acclaimed initiative:** Nursamsu et al., "WWF & Eyes on the Forest," Google Earth Outreach, https://www.google.com/earth/outreach/success-stories/wwf-eyes-on-the-forest.

175 **Another project making use:** Jim Robbins, "With an Internet of Animals, Scientists Aim to Track and Save Wildlife," *New York Times*, June 9, 2020, https://www.nytimes.com/2020/06/09/science/space-station-wildlife.html.

175 **One is the creation:** Donna Lu, "Two-Sided Solar Panels That Track the Sun Produce a Third More Energy," *New Scientist*, June 3, 2020, https://www.newscientist.com/article/2245180-two-sided-solar-panels-that-track-the-sun-produce-a-third-more-energy.

175 **Even more potentially transformative:** Julia Pyper, "How Innovator Bill Gross's Solar Breakthrough Could Decarbonize Heavy Industry," *GreenBiz*, March 4, 2020, https://www.greenbiz.com/article/how-innovator-bill-grosss-solar-breakthrough-could-decarbonize-heavy-industry.

176 **One last hopeful development:** Mahdokht Shaibani, "Solar Panel Recycling: Turning Ticking Time Bombs into Opportunities," *PV Magazine*, May 27, 2020, https://www.pv-magazine.com/2020/05/27

/solar-panel-recycling-turning-ticking-time-bombs-into-oppor
tunities.

176 **Some technology experts:** Seth Porges, "The Futurist: Why the
iPhone Reeks of Planned Obsolescence," *TechCrunch*, June 14, 2007,
https://techcrunch.com/2007/06/14/the-futurist-why-the-iphone
-reeks-of-planned-obsolescence.

177 **The preowned market:** Hannah Lutz, "Used-Vehicle Sales Drive
Dealership," *Automotive News*, November 12, 2018, https://www.au
tonews.com/article/20181112/RETAIL04/181119977/used-vehicle
-sales-drive-dealership.

177 **In 1955, the company's:** Don Hammonds, "Harley Earl's Design Innovations Changed the Auto Industry," *Pittsburgh Post-Gazette*, October 16, 2008, https://www.post-gazette.com/business/businessnews
/2008/10/16/Harley-Earl-s-design-innovations-changed-the-auto
-industry/stories/200810160525.

177 **Founder Bas van Abel:** Bas van Abel, "Fairphone—Changing the
Way Products Are Made," video, TEDx Amsterdam, November 6,
2013, https://www.youtube.com/watch?v=96XfmrJMlNU.

178 **Van Abel crafted:** Michael D'heur, *Sustainable Value Chain Management* (New York: Springer, 2015), 127.

178 **A website featuring:** Van Abel, "Fairphone—Changing the Way
Products Are Made."

178 **In June 2020:** Damien Wilde, "2015's 'Sustainable' Fairphone 2 Is
Now Receiving the Android Pie Update," 9to5Google, June 16, 2020,
https://9to5google.com/2020/06/16/fairphone-2-android-pie.

179 **Modularity may need:** Kurt Wagner, "Facebook Just Bought a Small
Hardware Startup Called Nascent Objects," *Vox*, September 19, 2016,
https://www.vox.com/2016/9/19/12974572/facebook-nascent-objects
-acquisition.

179 **One day in early 2020:** Paul Roberts, "Project BioMed—The Fight to
Repair Medical Devices," *Security Ledger* podcast, episode 184, May 13,
2020, https://securityledger.com/2020/05/episode-184-project-biomed
-the-fight-to-repair-medical-devices.

179 **But biomeds are:** *FDA Report on the Quality, Safety, and Effectiveness
of Servicing of Medical Devices*, 2018, i, https://www.fda.gov/media
/113431/download.

180 **Biomed Nader Hammoud:** Author interview with Nader Hammoud,
June 8, 2020.

180 **Because, as University of Pittsburgh:** An-Li Herring, "'Right-to-
Repair' Advocates Worry That Hospitals Can't Fix Broken Ventilators," Pittsburgh National Public Radio, April 20, 2020, https://www
.wesa.fm/post/right-repair-advocates-worry-hospitals-cant-fix
-broken-ventilators#stream/0.

180 **I asked Nathan Proctor:** Author interview with Nathan Proctor, June 5, 2020.

180 **I recalled the 2018:** Minda Zetlin, "The 9 Weirdest and Most Hilarious Questions Congress Asked Mark Zuckerberg," *Inc.*, April 12, 2011, https://www.inc.com/minda-zetlin/mark-zuckerberg-congress -hearings-funny-stupid-questions.html.

181 **In Europe, the push:** "EU Parliament Calls for Longer Lifetime for Products," EU Business.com, n.d., https://www.eubusiness.com/news -eu/durable-products.47bf.

181 **In 2020, the EU:** Rosie Frost, "Laptop Getting Old, Phone Due for an Upgrade? Don't Replace It, Repair It," *Euronews*, August 5, 2020, https://www.euronews.com/living/2020/05/08/laptop-getting-old -phone-due-for-an-upgrade-don-t-replace-it-repair-it.

182 **The Security Innovation Center:** Nicholas Deleon, "Right-to-Repair Laws Could Make It Easier to Get a Phone or Laptop Fixed," *Consumer Reports*, March 29, 2018, https://www.consumerreports.org /consumer-protection/right-to-repair-laws-could-make-it-easier -to-get-a-phone-or-laptop-fixed/.

182 **Yet, as *Consumer Reports*:** Deleon, "Right-to-Repair Laws Could Make It Easier to Get a Phone or Laptop Fixed."

182 **I spoke with technology:** Author interview with Paul F. Roberts, June 5, 2020.

182 **Going a step:** "Samsung Begins 2020 by Winning Three Awards for Commitment to Sustainability," Samsung Newsroom, January 8, 2020, https://news.samsung.com/us/samsung-2020-winning-three-awards -commitment-sustainability/.

182 **Another leader is:** Chris Nuttall, "Giving Old Tech a Second Life," *Financial Times*, October 5, 2019, https://www.ft.com/content/990c7846 -e5cf-11e9-9743-db5a370481bc.

Chapter 9: Building to Heal

184 **Monterey Bay National Marine:** Ross Clark, "Earth Matters: Researching the Corals of Monterey Bay Sanctuary," *Santa Cruz Sentinel*, October 4, 2018.

185 **One breed of:** Richa Malhotra, "The 16 Most Amazing Nests Built by Birds," Earth, BBC.com, March 7, 2015, https://www.bbc.com/earth /story/20150307-the-16-most-amazing-bird-nests.

185 **Naturalist Bernd Heinrich:** Bernd Heinrich, "African Social Weaverbirds Take Communal Living to a Whole New Level," *Audubon*, March–April 2014, https://www.audubon.org/magazine/march-april -2014/africas-social-weaverbirds-take-communal.

186 **The manufacture of cement:** Elisheva Mittelman, "The Cement Industry, One of the World's Largest CO_2 Emitters, Pledges to Cut Greenhouse Gases," YaleEnvironment360, Yale School of the Envi-

ronment, December 28, 2018, http://e360.yale.edu/digest/the-cement
-industry-one-of-the-worlds-largest-co2-emitters-pledges-to-cut
-greenhouse-gases.

186 **In fact, it's said:** Lucy Rodgers, "Climate Change: The Massive CO_2 Emitter You May Not Know About," BBC News, December 17, 2018, https://www.bbc.com/news/science-environment-46455844.

187 **Constantz says the partnership:** PR NewsWire, "Blue Planet and KDC Announce a Global Partnership to Scale Carbon Negative Concrete Worldwide—Solving One of the Largest Environmental Problems," press release, December 5, 2019, https://apnews.com/press-release/pr-prnewswire/1ea2323684f5528c3b58775acaa25a53.

187 **At the University of Colorado Boulder:** Amos Zeeberg, "Bricks Alive! Scientists Create Living Concrete," *New York Times*, January 15, 2020, https://www.nytimes.com/2020/01/15/science/construction-concrete-bacteria-photosynthesis.html.

187 **Growing and restoring:** Huntley Penniman, "Coral Reefs and Climate Change: What to Know and What to Do," Oceanic Society, http://www.oceanicsociety.org/blog/2201/coral-reefs-and-climate-change-what-to-know-and-what-to-do.

187 **As reefs support:** Penniman, "Coral Reefs and Climate Change."

188 **The iron-and-wood framing:** Elizabeth Hone, "The Gardener Who Nurtured London's Crystal Palace," *Christian Science Monitor*, October 24, 1980, https://www.csmonitor.com/1980/1024/102456.html.

188 **The interior of:** "The Interior of the Sagrada Familia," Barcelona.de, https://www.barcelona.de/en/barcelona-sagrada-familia-interior.html.

188 **the ancient Roman:** "Roman Aqueducts," Resource Library, *National Geographic*, https://www.nationalgeographic.org/encyclopedia/roman-aqueducts/.

188 **On Sunday, October 8:** "The Great Chicago Fire of 1871," Chicago Architecture Center, https://www.architecture.org/learn/resources/architecture-dictionary/entry/the-great-chicago-fire-of-1871.

189 **When Chicagoans rebuilt:** Donald L. Miller, *City of the Century: The Epic of Chicago and the Making of America* (New York: Rosetta Books, 2014), 306.

189 **As Neal Bascomb describes:** Neal Bascomb, *Higher: A Historic Race to the Sky and the Making of a City* (New York: Broadway Books, 2003), 97.

189 **historian Donald L. Miller:** Miller, *City of the Century*, 306.

189 **Miller also quotes:** Miller, *City of the Century*, 306.

190 **In Miller's words:** Miller, *City of the Century*, 307.

190 **The result of the relentless:** Guglielmo Carra and Arup Nitesh Magdani, "Circular Business Models for the Built Environment," Ellen MacArthur Foundation, 3.

190 **Brent Constantz cautioned:** Don Procter, "Construction Concrete a Key 'Lever' in Climate Change Fight: Expert," October 6, 2019, *Daily Commercial News*, https://canada.constructconnect.com/dcn/news /others/2019/10/construction-concrete-a-key-lever-in-climate-change -fight-expert.

190 **So rapid is:** Bruce King, *The New Carbon Architecture: Building to Cool the Climate* (Gabriola, BC: New Society Publishers, 2017), 11.

191 **Paul and his son:** Author interview with Paul and Kyle Macht, July 1, 2020.

191 **This fundamental wisdom:** Dennis Holloway, "Sun Tempered Architecture," Dennis R. Holloway, Architect, http://www.dennisrhol lowayarchitect.com/SimpleDesignMethodology.html.

193 **A study conducted:** G. Churkina et al., "Buildings as a Global Carbon Sink," *Nature Sustainability* 3 (2020): 269–76, https://doi.org/10 .1038/s41893-019-0462-4.

193 **An 18-story, 280-foot:** India Block, "Mjøstårnet in Norway Becomes World's Tallest Timber Tower," *Dezeen*, March 19, 2019, https://www .dezeen.com/2019/03/19/mjostarne-worlds-tallest-timber-tower-voll -arkitekter-norway.

193 **Numerous plans have:** Tom Ravenscroft, "World's Tallest Timber Tower Proposed for Tokyo," *Dezeen*, February 19, 2018, https://www .dezeen.com/2018/02/19/sumitomo-forestry-w350-worlds-tallest -wooden-skyscraper-conceptual-architecture-tokyo-japan.

194 **In a nod to biomimicry:** "Living Building Challenge," International Living Building Institute, https://living-future.org/lbc-3_1.

195 **I spoke to McLennan:** Author interview with Jason McLennan, June 30, 2020.

195 **"Nature—not ego":** Jason McLennan, *Zugunruhe: The Inner Migration to Profound Environmental Change* (Bainbridge Island, WA: Ecotone Publishing, 2011), 47.

196 **Berkebile and a group:** Brian J. Barth, "The Past, Present, and Future of Sustainable Architecture," *Pacific Standard*, June 13, 2018, https:// psmag.com/environment/past-present-and-future-of-sustainable -architecture.

196 **He found that:** Jill Fehrenbacher, "Biomimetic Architecture: Green Building in Zimbabwe Modeled After Termite Mounds," *Inhabitat*, November 29, 2012, https://inhabitat.com/building-modelled-on-termites -eastgate-centre-in-zimbabwe.

197 **Mimicking this system:** "Passive and Low-Energy Heating and Cooling Saves Building Costs," Ask Nature, October 1, 2016, https://askna ture.org/idea/eastgate-centre/.

197 **The result, as one:** Rochelle Ade and Michael Rehm, "The Unwritten History of Green Building Rating Tools: A Personal View from Some

of the 'Founding Fathers,'" *Building Research & Information* 48, no. 1 (2020): 1–17, doi: 10.1080/09613218.2019.1627179.

197 **"The certification has become":** Ade and Rehm, "The Unwritten History of Green Building Rating Tools."

198 **"Every time industry":** Jason McLennan interview.

198 **The Bullitt Center:** "The Greenest Commercial Building in the World," Bullitt Center, https://bullittcenter.org.

198 **The Kendeda Building:** "Celebrating the Kendeda Building," Kendeda Foundation, https://livingbuilding.kendedafund.org/2020/06/09/salvaged-materials-tell-buildings-best-stories.

198 **McLennan sums up:** "A Case for Regenerative Design: An Interview with Jason McLennan," Design Intelligence, November 6, 2018, https://www.di.net/articles/case-regenerative-design-interview-jason-f-mclennan/.

Chapter 10: Scaling Circularity Up

201 **Recall that Hollender:** Author interviews with Jeffrey Hollender, June 10 and June 17, 2020.

202 **Sales of consumer goods:** Randi Kronthal-Sacco and Tensie Whelan, "Sustainable Market Share Index: Research on 2015-2020 IRI Purchasing Data Reveals Sustainability Drives Growth, Survives the Pandemic," New York University Stern School of Business, updated July 16, 2020, https://www.stern.nyu.edu/sites/default/files/assets/documents/NYU%20Stern%20CSB%20Sustainable%20Market%20Share%20Index%202020.pdf.

204 **During his ten years:** Details and quotes not otherwise cited based on author interview with Paul Polman, June 21, 2020.

204 **In announcing the plan:** Dan Schawbel, "Unilever's Paul Polman: Why Today's Leaders Need to Commit to a Purpose," *Forbes*, November 21, 2017.

204 **In 2019, Unilever:** Unilever, "Unilever's Purpose-Led Brands Outperform," press release, November 6, 2019, https://www.unilever.com/news/press-releases/2019/unilevers-purpose-led-brands-outperform.html.

205 **The company also:** "Making Sustainable Living Commonplace for 8 Billion People," Unilever, 2020, https://www.unilever.com/sustainable-living/ten-years-on.

206 **The World Economic Forum:** World Economic Forum, *An Economic Opportunity Worth Billions—Charting the New Territory*, https://reports.weforum.org/toward-the-circular-economy-accelerating-the-scale-up-across-global-supply-chains/an-economic-opportunity-worth-billions-charting-the-new-territory.

207 **A McKinsey analysis concludes:** Clarisse Magnin and Saskia Hedrich, "Refashioning Clothing's Environmental Impact," McKinsey &

Company, https://www.mckinsey.com/business-functions/sustainabi lity/our-insights/sustainability-blog/refashioning-clothings-envi ronmental-impact#.

207 **Recall that the overall:** Peter Lacy and Jakob Rutqvist, *Waste to Wealth: The Circular Economy Advantage* (New York and London: Palgrave Macmillan, 2015).

207 **A Harvard study:** Robert G. Eccles, Ioannis Ioannou, and George Serafeim, "The Impact of Corporate Sustainability on Organizational Processes and Performance," *Management Science*, November 2014, https://www.hbs.edu/faculty/Publication%20Files/SSRN-id1964011 _6791edac-7daa-4603-a220-4a0c6c7a3f7a.pdf.

207 **Research by financial assets:** Matt Heimer, "Doing Well by Doing Good: 5 Stocks to Buy for 2019," *Fortune*, December 5, 2018, https:// fortune.com/2018/12/05/best-stocks-esg-2019-walmart-abbott-merck.

208 **Growing up as a track star:** Author interview with Ginger Spencer, June 24, 2020.

209 **In England, the political:** "Circular Peterborough," FuturePeterbor ough.com, http://www.futurepeterborough.com/circular-city/.

210 **In the U.S., leaders in Charlotte:** "What Is a Circular Economy," En vision Charlotte, https://envisioncharlotte.com/circular-charlotte.

210 **In response to the COVID-19 pandemic:** Ewa Krukowska and Laura Millan Lombrana, "EU Approves Biggest Green Stimulus in History with $572 Billion Plan," *Bloomberg*, July 21, 2020, https://www.bloom berg.com/news/articles/2020-07-21/eu-approves-biggest-green-stim ulus-in-history-with-572-billion-plan.

211 **China has also announced:** Naomi Xu Elegant, "China's Ambitious New Climate Goals Mean It Must Completely Phase Out Coal by 2050," *Fortune*, September 28, 2020, https://fortune.com/2020/09/28 /china-climate-change-goals-coal.

211 **In the U.S., in addition:** Megan Smalley, "Federal Legislators Share Break Free from Plastic Pollution Act Blueprint with State Legisla tors," *Recycling Today*, August 12, 2020, https://www.recyclingtoday .com/article/senators-share-blueprint-break-free-plastic-pollution -act-lawmakers.

Printed in the United States
by Baker & Taylor Publisher Services